the COMPASS

How to Navigate AI for the Non-Technologist

BOB PEARSON

author of *Storytizing*

the COMPASS

How to Navigate AI *for the* Non-Technologist

BOB PEARSON
author of *Storytizing*

1845 PUBLISHING

the
COMPASS

by Bob Pearson

ISBN: 978-1-7377964-3-5

Cover and interior design by: TLC Book Design, *TLCBookDesign.com.*

Cover: Tamara Dever / Interior: Erin Stark

Compass graphic: *Vecteezy.com*

Printed in the United States of America.

*Thank you to the leaders who
taught me how to "think different,"
and for my family who believes
in creative expression of ideas.*

Table of Contents

PART THREE
Questions, Context, and Decision Quality

PART FOUR
Media, Creativity, and Communication in an AI World

PART FIVE
Security, Ethics, and Governance

PART SIX
Strategy, Leadership, and the Next Decade

Introduction

I didn't plan to write a trilogy—it emerged from being repeatedly placed on the front edge of technological change.

Sometimes history hands you a front row seat, and you either take notes or get swept away.

If you go back to the beginning of this century, we were just starting to experience an explosion of social media platforms: Friendster (2002), LinkedIn and MySpace (2003), Facebook (2004), YouTube (2005), Twitter (2006), Tumblr (2007), WhatsApp and Pinterest (2009), and more—all riding on the success of Google, which launched in 1998.

At Dell Technologies, we built one of the corporate world's first global social media functions. Our team launched the first corporate blog in China. We ran more than one hundred corporate Twitter accounts. I remember joining early conversations with Facebook about international expansion and visiting YouTube to understand how they planned to work with brands.

Amidst all of that innovation, one moment stands out: In 2007, Dell's first blog leader, Lionel Menchaca, called me from SXSW and said, "There's this new thing called Twitter. Think we should use it?" My answer: "Never heard of it, but yes, let's try." After SXSW, daily tweets jumped from about twenty thousand to sixty thousand. Today, there are roughly five hundred million posts per day.

Back then, everything felt like it was changing—business models, how we marketed, where customers gathered, who had the power to shape a narrative. That period inspired my first book, *Pre-Commerce* (2011).

Pre-Commerce made a simple argument: Social media was fundamentally changing how customers learn, decide to buy, and stay loyal. Most of the real action happens **before** or **after** a transaction, not in the moment of purchase. If you could understand and influence that broader journey, you would win.

At the time, this felt like a provocative idea. Many leaders quietly understood the shift, but systems, incentives, and habits were still aligned with the old world. Over time, what we once called "pre-commerce" just became **commerce**.

After leaving Dell, I joined WeissComm, which became W2O Group and now Real Chemistry, a leading healthcare agency. There, I began to see that social media was only one part of a larger shift toward data-driven storytelling—how we use data to understand audiences, craft narratives, and build lasting relationships.

That led to my second book, *Storytizing* (2016). The premise was that the traditional advertising model—one-way messages pushed at mass audiences—was becoming obsolete. Instead, brands and customers would co-create stories. Audience architecture, micro-segmentation, and analytics would reshape how we decide what to say, where to say it, and to whom.

Storytizing forecast an era where content creators, influencers, and communities would hold more power than traditional media. When the book came out, I had many private conversations with leaders who admitted the ad model was changing forever—just not publicly. Today, brands actively court influencers and creators to co-build relevance. The prediction became reality.

My goal in writing these books has always been the same:

- Help us all look around the corner.
- Imagine what is possible.
- Recognize that clinging to what works now can be the riskiest strategy of all.

Interestingly, the feedback loop is delayed. I tend to hear more from readers three to five years after a book comes out than in the first three to five months. That's fine. These ideas are meant to age into relevance, not just spike attention.

After *Storytizing*, I wasn't sure what the "third act" would be. Then, in late 2022, generative AI surged into the mainstream and changed the conversation almost overnight.

This time felt different. The hype was—and is—overwhelming. It's understandable why:

- More than **five billion** people are online and will be touched by AI.
- Tens of thousands of AI-focused companies are founded each year.
- New tools appear weekly that claim to transform how we work.

As with past waves, the winners will be those who apply technology to solve real problems and create real opportunities—especially "last-mile" solutions and ecosystem-level improvements.

But there's a major difference this time, and that's why I wrote *The Compass*:

- **Everyone** can benefit from AI.
- **Everyone** can also be left behind by AI.

AI will create new business models and destroy old ones. It can be deeply human-centered, or it can be used in ways that amplify bias and erode trust.

What we need is a way to navigate the uncertainty that AI creates, which led me to think of a compass.

A compass helps us understand where we are, no matter what the conditions. It doesn't eliminate uncertainty, but it gives us a stable point of reference. It tells us which direction is north so we can decide where to go next.

We are entering a period where we will experience more change in the next twenty years than we did in the last one hundred. To succeed, we must learn how to **think alongside technology**, not chase every new tool or ignore it altogether. We need to guide AI, not be dragged by it.

And as a result, we need to know where we should spend our time.

That is the purpose of this book.

The Compass is about helping you:

- Build your own "AI compass."
- Use it to navigate opportunities and risks.
- Make better decisions for yourself, your team, your organization, and society.
- Avoid following the wrong signals into storms you could have avoided.

In navigation, we distinguish between **true north** and **magnetic north**.

- True north points to the North Pole.
- Magnetic north is where the compass needle actually points based on the Earth's magnetic field, which shifts over time.

AI is similar. There will always be a "true north" for your organization—your mission, your values, what kind of impact you want to have. But the "magnetic north" of technology will keep shifting as models evolve, regulations change, and new capabilities emerge.

This book will help you do both:

- Stay anchored to your true north as a leader
- Learn to read the magnetic field of AI so you can move with confidence

Best,

Bob Pearson

PART ONE

Becoming a Navigator in the Age of AI

1

Navigating the Potential of AI

"The only thing we have to fear is fear itself."
– Franklin D. Roosevelt

Your role as a leader is to navigate thoughtfully, not predict everything. Your advantage won't derive from knowing every detail. It will come from how you balance curiosity, self-awareness, and deliberate experimentation.

Imagine living in the 1760s, and someone named James Watt tells you he's invented a machine that can replace manual labor, power water mills, run factories, and move trains and ships. He calls it a steam engine.

In 1765, many would have thought he was mad.

He wasn't. The steam engine helped unlock a new era in manufacturing, transportation, and trade. It reshaped where people lived, how they worked, and how wealth was created.

Large transformations are always unsettling and inspiring at the same time. They excite founders and innovators. They unsettle incumbents and those whose routines are disrupted.

In every era, we are told, "This will change everything." Most things don't. Some things absolutely do.

Our challenge is that our brains are wired for predictability. Evolution teaches us to notice threats more than opportunities. A small almond-shaped structure in our brain—the amygdala—processes emotions and tags memories with emotional weight.

When something big and unfamiliar appears, the amygdala often leads with: "Threat. Danger."

From there, our insecurities fill in the story:

- "Will I still have a job?"
- "Are my competitors getting ahead of me?"
- "Am I too far behind to catch up?"
- "Do I really have to learn something completely new again?"

We even have sayings to comfort ourselves: "Better the devil you know than the devil you don't."

But in times of major transformation, that mindset can quietly block progress.

A better way is to **slow the story down**:

- Write down what could go right.
- Write down what could go wrong.
- Think methodically about possibilities and risks.

When we approach change this way, we move from vague anxiety to concrete choices. We start to see paths that are:

- ambitious and realistic
- innovative and responsible

Real transformation takes all of us, not just technologists.

Think about the printing press. Before it, books and documents were written by hand. They were slow to produce, hard to distribute, and limited to small groups.

In the 1440s, Johannes Gutenberg developed a printing press with movable type—a breakthrough, but not the whole story. Other leaders had to:

- build more presses
- improve the mechanics (like cast iron and steam power)
- create systems for distribution, education, and publishing

A generation of innovators took a powerful idea and **stretched it step by step**.

The same pattern repeats in modern times.

- Tim Berners-Lee created the World Wide Web in 1989.
- Google launched in 1998.
- Facebook arrived in 2004, Spotify in 2008, TikTok in 2017.
- The Human Genome Project began in 1990, with a draft genome announced in 2001.

We get excited about a concept, capital flows in, headlines explode—and then the slow, compounding work of real integration begins.

That is where leaders earn their reputation.

We are now in a similar moment with AI. Headlines make it sound like it all happened overnight. In reality, what you see today is the product of decades of groundwork in:

- computing hardware
- data storage and infrastructure
- networking
- algorithms and open-source software
- venture-backed risk-taking

Your role is not to become an AI scientist. Your role is to become a **navigator of change**.

The most fundamental tool of a sailor is their compass, which is designed to work in storms and calm weather alike. It doesn't remove uncertainty. It gives you a reliable orientation so you can decide:

- when to move
- where to go
- what to avoid

The goal of this book is to help you build your own compass for AI so you can chart a course that is bold, responsible, and grounded in reality.

2

Why AI and Humans Need Each Other

AI is at its best when it amplifies distinctly human strengths—judgment, ethics, empathy, creativity, and context. Our job is not to choose between humans and AI, but to redesign work so they reinforce each other.

Let's start by familiarizing ourselves with generative AI.

It is different in one important way: It doesn't just move things or store things. It **creates** things—text, images, code, even video. So it can feel, at first glance, like it's doing "our" kind of work.

Here's the reality:

AI is extraordinarily good at pattern recognition, prediction, and scale.

Humans are uniquely good at meaning-making, ethics, context, and care.

When we frame it this way, the question shifts from "Will AI replace us?" to "What are our best roles?"

"Where do we want machines to be in charge, and where must humans stay in the loop?"

- You want AI to rapidly analyze millions of data points.
- You want humans to decide how to optimize that data.
- You want AI to propose scenarios.

- You want humans to weigh trade-offs.
- You want AI to help draft content.
- You want humans to own the narrative and the truth.

In practical terms:

- AI can read every customer ticket; humans decide what policy to change.
- AI can scan every contract; humans decide what risk is acceptable.
- AI can generate a hundred creative concepts; humans decide what fits the brand and the moment.

If you try to remove humans entirely, you lose judgment, accountability, and trust. If you ignore AI, you sacrifice speed, precision, and scale.

The organizations that will win are the ones that learn how to design human–AI workflows, not human or AI workflows.

That's why the rest of this book focuses on:

- how humans think and build habits
- how AI actually works (without jargon)
- how to redesign roles, workflows, and incentives
- how to keep your organization's values at the center

AI needs humans. Humans now need AI. The leadership opportunity is to make that partnership intentional, not accidental.

3

Why Humans Are Necessary to Drive AI Solutions

AI is extraordinary at logic, pattern recognition, and processing the past. But it lacks "non-logical intelligence"—story, emotion, intuition, and lived context. That makes humans indispensable as navigators, designers, and imaginative partners in the AI era. Human judgment remains essential.

Common Sense Is Still Hard for Machines

Hector Levesque, professor emeritus at the University of Toronto, proposed the Winograd Schema Challenge to test whether machines can handle common-sense reasoning.

Example:

> "The city councilmen refused the demonstrators a permit because they feared violence."

Who feared violence? ⇨ The **council members**.

Change one word:

> "The city councilmen refused the demonstrators a permit because they **advocated** violence."

Now "they" refers to the demonstrators.

Humans navigate this effortlessly using:

- context
- world knowledge
- social understanding

Machines can be trained to perform well on tests like this, but it highlights a deeper point:

AI is brilliant at logic and patterns, not at fully inhabiting our messy, human world.

What AI Is Exceptionally Good At

- processing vast amounts of data from the past
- applying math and probability at massive scale
- recognizing patterns we can't see
- optimizing known systems

If something is structured, historical or quantitative, AI is likely to outperform us.

But predicting the future with certainty?

- Daily stock prices with 100 percent accuracy?
- Exact pricing for every item over the next year?

That's not going to happen—and not just because of technical limits. The world is shaped by *human behavior* that doesn't follow pure logic.

Non-Logical Intelligence — Our Human Advantage

What machines still lack is what we might call non-logical intelligence:

- Language and storytelling
  - understanding tone, subtext, irony, humor
- Sound and music
  - feeling what a certain chord progression communicates emotionally
- Visual thinking
  - imagining a world that doesn't yet exist from a rough sketch or metaphor
- Emotional intelligence
  - reading a room through micro-expressions and atmosphere
- Risk and danger sensing
  - intuitively noticing when something "feels off" before we can explain why
- Pattern intuition
  - catching emergent patterns in culture, markets, or behavior that no model has fully captured yet

This is our world:

- We understand unspoken rules.
- We sit with uncertainty.
- We imagine futures that are not yet in any dataset.

We also created AI in the first place.

Humans as Navigators of AI

When we accept this division of strengths:

- AI: logic, pattern, past
- Humans: context, emotion, imagination, future

...we stop trying to compete with the machine at what it does best and instead ask:

How do we design applications that combine AI's brilliance with our own?

A few examples:

- A surgeon uses AI to simulate multiple surgical paths but chooses based on experience and nuance.
- A farmer uses AI to analyze soil, climate, and yield, but decides based on local knowledge and risk appetite.
- A trader uses AI to surface signals but draws on pattern intuition and judgment to act.

There is plenty for both of us to do.

Airbnb: A Human Idea, AI-Scalable

Homeowners have always had spare rooms.

It took Airbnb to turn that into a global business:

- The founders rented out their own apartment in 2007 during a conference to cover rent.
- They noticed an emotional truth: People like the idea of feeling "at home" anywhere in the world.

- The insight was not about *cheaper* or *more convenient* lodging. It was about belonging, story, and experience.

Only humans could have:

- felt that emotional resonance
- framed the value proposition that way
- persisted through the early skepticism

AI can help:

- match guests and hosts
- optimize pricing
- detect fraud
- personalize recommendations

But the **idea** itself was human.

4

The Non-Technologist Is the Compass

You know the customer, their pain points, and the opportunities. It is why you are the compass for AI and those who build future AI apps. Our job is to know the problems, challenge assumptions, and partner with technologists as equals.

Our brains love routines. We like getting into a groove. So when something like generative AI hits our shores, it can feel like a tsunami has arrived—and many of us were quite happy surfing gentle three-to-five-foot waves.

We often make AI more complicated than it needs to be. At its core, innovation is still what it has always been: **solving problems that matter to people**.

If you look at how inventors are defined, the essence is simple:

Someone identifies an unmet need or a problem and then figures out a solution.

Notice what's missing from that definition:

- It doesn't say you need to know how to code.
- It doesn't say you must master every AI model.
- It doesn't require a zip code in Silicon Valley.

Most of the world's problems are understood first by **non-technologists**:

- clinicians at the bedside
- sales teams in the field
- operations managers in plants and warehouses
- customer service teams listening to complaints
- marketers seeing where trust is eroding

These are the people who see reality up close. They know where things are breaking, where customers are frustrated, where employees are stuck.

Think of generative AI as a magnetic field and think of yourself as the arrow.

The magnetic field is powerful and constant. It doesn't care about your org chart. It will keep reshaping markets and expectations whether you engage with it or not.

But the arrow—your judgment, your choice of direction—is what determines where you actually go.

Non-technologists have three critical roles:

1. Define the problem worth solving.

- "What would it mean to cut onboarding time in half?"
- "What would it mean if every rep walked into a meeting fully briefed?"
- "What if a rare-disease patient never had to repeat their story?"

2. Partner with those who can build.

- You don't have to write the code.
- You do need to sit side by side with technologists and say, "That solves the wrong problem," or "Here's the nuance you're missing."

3. Guardrails and relevance.

- You know what is ethical or tone-deaf in your industry.
- You know what would delight a customer, and what would creep them out.

The biggest barrier is often not capability—it's self-confidence.

We tell ourselves:

- "I'm not technical enough."
- "I don't understand the models."
- "My ideas probably aren't that original."

In reality, many of the best AI solutions will come from unlikely pairings: an operations manager with a pattern-recognition itch plus an engineer who knows how to build agents.

It's like peanut butter and chocolate: good on their own, much more interesting together.

If you are a non-technologist, consider this mindset shift:

- You are not a passenger in the AI era.
- You are the compass—the one who points builders toward real problems that matter.

Every pilot program that you conduct, whether it succeeds or fails, teaches you something. Over time, those learnings become your map. And your willingness to raise your hand and say, "There's a better way to do this," is often the spark that AI needs.

5

How to Study Technology's Advance

If you consistently ask how a technology expands capacity, speeds up information, scales impact, changes work, and creates new markets, you will see the important patterns early enough to react.

The pace of technology can feel mind-numbing. Every day brings new announcements, new acronyms, and new promises. It's easy to feel that keeping up is a full-time job.

At the same time, history shows that real adoption takes years—even decades. Headlines move fast; infrastructure and behavior move slower.

To avoid being overwhelmed, I use a simple strategy to study technology's advance. I ask five questions:

1. Will this technology expand capacity?
2. Will it accelerate information processing?
3. Does it represent a breakthrough in how to scale?
4. Will it change how we work?
5. Will it create entirely new market opportunities?

Let's take them one at a time.

1. Will this technology expand capacity?

The cloud is a good example. Initially, it was about storage and flexibility. Now it's becoming a way to rent intelligence (e.g., our large language models), not just store information. We are moving toward what you might call Cloud 2.0—where capacity isn't just about disk space, but about models, agents, and decision engines.

The strategic question becomes:

Who is expanding capacity in a way that unlocks better decisions and better outcomes?

For example, think about network infrastructure. 5G may feel a bit underwhelming, but 6G is projected to be at least one hundred times faster. It will support billions of devices, agents, wearables, and immersive experiences. What kind of capacity will that require?

Countries and companies that underinvest in this infrastructure will quietly fall behind. Those who invest intelligently will have a richer canvas for AI.

2. Will this technology accelerate information processing?

We are used to computers doing one thing at a time very quickly. Quantum computing flips that idea—processing many possibilities simultaneously.

The question for leaders is not, "How does quantum computing work?" The practical questions are:

- Where will its impact be felt first?
  - Drug discovery?
  - Financial modeling?
  - Climate simulations?
  - Security and cryptography?
- How would your industry change if some problems could be solved in minutes instead of months?

You don't need to understand every math detail. You need a sense of **where the acceleration of technology will hit your market or create a new one.**

3. Does this technology represent a breakthrough in how to scale?

Voice and language technologies are a good example. Tools that can translate your speech into dozens of languages almost instantly are not just "cool features." They change the economics of:

- customer service
- media distribution
- product launches
- training and education

One person can now communicate worldwide in any language or dialect at essentially zero marginal cost. That's a scaling breakthrough.

The question becomes:

Who will use this leap to reimagine how they reach and support people?

4. Will this technology change how we work?

Agents—software entities that can observe, decide, and act—will increasingly become part of our daily workflow.

They will:

- prepare meeting briefs
- summarize documents
- draft responses
- monitor signals
- triage issues

The more a firm embraces these digital coworkers, the more its work capacity changes. The interesting question is not, "Will agents exist?" They already do.

The questions are:

- What kind of work do you want them to do?
- How do you train and govern them?
- How does your org chart evolve when some "roles" can be performed by AI?

5. Will a new market emerge because of this technology?

Sometimes technology doesn't just improve existing markets. It creates new ones.

For example:

- A space-based economy is emerging—with satellites that extend the internet, monitor climate, and support navigation.

- CRISPR is enabling gene editing that could transform medicine and agriculture.

You don't need to bet on every frontier, but it's helpful to keep a list of "watch areas" where entirely new industries may form.

Thinking Way Ahead

Here's how this way of thinking might show up in everyday scenarios:

- **Cloud** – You can imagine combining all of the health information for every patient for their entire life in a secure setting that the patient and physician can access.
- **6G** – You meet a colleague in France in a small room in Austin using full-body holograms. It feels like a normal meeting—no travel needed.
- **Quantum** – A tsunami hits and shipping routes are recalculated in seconds.
- **Voice** – You record a thirty-minute podcast; it's instantly translated into fifty languages and distributed while you get coffee.
- **Agents** – One person launches a global brand using agents for design, pricing, logistics, and customer service.
- **Space** – A terrestrial grid fails, and space-based networks keep hospitals and schools online.
- **CRISPR** – Doctors remove a defective gene before birth as part of routine care.

The point isn't to predict every scenario. It's to discipline your curiosity.

Rather than chasing every headline, keep asking:

Will this expand capacity, speed up information, change how we scale, reshape work, or create new markets?

When you do, you'll find it easier to decide where to lean in, where to wait, and where to say, "No, not relevant...yet."

6

The Journey of AI Itself

AI is built on seventy-five years of progress in logic, computing, data, networking, software, and capital. It is a tool we can shape, not a force of nature we have to accept.

AI didn't just "happen." What you see today is the product of many layers of innovation coming together over time.

If this were a sports highlight show, here's the short reel:

- **19th century** – George Boole introduces Boolean logic. Charles Babbage and Ada Lovelace sketch out ideas that foreshadow programmable machines.
- **Mid-20th century** – Alan Turing asks, "Can machines think?" and proposes the Turing Test—a way of seeing whether a machine can imitate human conversation well enough to fool a human judge.
- **Post-war computing** – ENIAC and UNIVAC appear, proving that programmable machines can solve real problems.
- **Transistors and microprocessors** – replaces vacuum tubes, dramatically increasing computing power while shrinking size and cost.
- **Moore's Law** – ushered in an era where transistor density doubles every couple of years.
- **Data explodes** –machine learning shifts from an academic idea to practical necessity.

- **Databases evolve** – from centralized systems to distributed frameworks like Hadoop, Spark, and NoSQL, making it possible to store and process massive datasets.
- **High-speed internet** – enables real-time sharing of data, content, and transactions.
- **Software ecosystems bloom** – from Java, C++, and Python to CUDA and specialized libraries.
- **Capital flows in** – funding ambitious experiments and companies.
- **Open-source communities** – share tools and models, accelerating progress.

By the time generative AI burst into the mainstream conversation in 2022, the foundation had been under construction for decades.

We also expanded how we measure "intelligence":

- The **Turing Test** focused on conversational imitation.
- The **Winograd Schema Challenge** tested common-sense reasoning.
- The **Lovelace Test** looked at whether a machine could produce something beyond what it was explicitly programmed to do.
- New tests now probe multimodal reasoning, exam performance, coding ability, and more.

It's tempting to think that whatever is happening right now is the final form. It isn't. We are standing on the shoulders of many innovators—and future leaders will stand on ours.

Your job is to use what they have made possible in a way they would be proud of.

PART TWO

Habits, Behavior, and the Human Mind

7

Why It Takes Time to Change Our Habits

If your incentives and rewards are tied to your old way of working, your AI strategy will stall. To make AI stick, you must intentionally redesign habits at the workflow and compensation levels.

It's easy to send an all-hands email declaring that your company is embracing AI. It's even easier to give an inspiring speech about automation, growth, and transformation.

But nothing truly changes unless habits change.

To understand why, we need a quick tour of our brain.

The basal ganglia plays a key role in cognition, emotion, learning, motivation, and habit formation. Habits form through a simple loop:

1. **Cue** – a trigger that starts the behavior
2. **Routine** – the behavior itself
3. **Reward** – the benefit that reinforces the loop

Most leaders conceptually understand this. Where we struggle is doing the hard, unglamorous work of designing new habits.

Consider a consulting firm.

- A client calls and says, "We need a media plan for a new brand by next Friday."

- That's the **cue**.
- The **routine** is well established: assign two account executives, invest fifty hours of work, build the plan in the usual way.
- The **reward**: the client is happy, the team logs fifty billable hours, and everyone feels productive.

Now imagine we introduce AI.

- The same request comes in.
- The team uses an AI platform to generate a draft media plan in minutes.
- They spend a day refining it, aligning it with strategy, checking the details.
- They deliver it six days early, with twenty-five hours of internal time instead of fifty.

The outcome is better:

- higher margin
- faster delivery
- more comprehensive plan

Everyone wins, right?

Not necessarily.

Here's what happens if we *don't* change the system:

- The finance director notices that only 50 percent of "normal" time was billed.
- People start worrying: "If we keep doing this, will my billable hours drop? Will my job be at risk?"
- Leaders start to wonder, "Do we need this many staff if AI is doing more?"

- The implicit message becomes: "If you adopt AI, you might be making yourself redundant."

So the next time a similar request comes in, the team quietly returns to the old routine. The habit snaps back into place.

The problem isn't the staff. It's the **system**.

If we want to leverage AI, we have to redesign:

- the **workflow** – routine
- the **metrics** – what we measure
- the **rewards** – what we celebrate and pay for

Back to our consulting firm example. As leaders, we might:

1. Map the current habit loop.

- Cues: client requests, new business opportunities, internal asks.
- Routines: who does what, for how long, using which tools.
- Rewards: billable hours, perceived busyness, promotions tied to time spent.

2. Redesign the routine.

- We explicitly decide: "From now on, all media plans will start with our AI partner."
- We set a new expectation: "This task should take twenty-five hours, not fifty."

3. Redesign the reward.

- We increase the bonus pool tied to **effective AI use** and **impact**, not hours billed.
- People who adopt AI are more likely to see bonus upside and new growth opportunities.

- The time saved is reallocated to higher-value work: more clients, deeper strategy, new offerings.

Now the message is different:

"If you adopt AI responsibly, you can earn more, grow faster, and do more interesting work."

The margin still improves. The firm can scale with more discipline. Hiring becomes more strategic rather than reactive.

The one approach that almost guarantees failure is using AI as an excuse for layoffs. If people feel that AI adoption equals job loss, they will resist it—consciously or subconsciously. You have, in effect, made "reward" equal "termination."

The lesson:

- You don't erase old habits.
- You *replace* them with new ones—keeping the cues, redesigning the routines, and aligning the rewards.

Then again, the best examples often come from outside technology.

Take Febreze, launched in 1998 by Procter & Gamble. Initially, it was positioned as an odor-neutralizing spray. Technologically clever, but it struggled. Customers didn't see it as essential.

The breakthrough came when P&G repositioned Febreze as the **final step in a cleaning routine**:

- The cue: finishing cleaning.
- The routine: spray Febreze as a last touch.
- The reward: a fresh, satisfying scent.

They didn't try to replace cleaning itself. They embedded Febreze into an existing habit and made it the satisfying finishing move.

The same principle applies to AI in your organization. Understanding and redesigning habits is not soft stuff—it is how you unlock margin, productivity, and morale.

8

Addressing Dopamine

"Quality is not an act; it is a habit."
– Will Durant

Designing healthy, meaningful dopamine loops at work is now a leadership responsibility. Internal environments that are dull can't compete with external platforms that are rewarding us.

In the previous chapter, we explored habits. To round out that picture, we need to talk about a key player in your brain's reward system: dopamine.

Dopamine is a neurotransmitter deeply involved in how we experience motivation and reward. It isn't just about pleasure. It's about anticipation, learning, and reinforcement.

Arvid Carlsson, a Swedish neuroscientist and Nobel laureate, helped show how dopamine is related to movement and conditions like Parkinson's disease. When certain brain cells that produce dopamine die, people can develop movement disorders—tremors, slowed motion, freezing.

Drugs like levodopa can temporarily alleviate symptoms by increasing dopamine, but they don't stop the underlying cell loss.

Why does this matter to leaders in the AI era?

Because the same system that helps us move and form habits is also engaged when we use technology.

When social platforms like Facebook and others introduced likes, comments, shares, reactions, and saves, they weren't just adding features. They were tapping into dopamine-driven loops.

Every notification, every small social reward, becomes a tiny hit of "this is interesting, come back."

We now live in a world where:

- Average time on mobile devices is measured in **hours per day**.
- We check our phones dozens (or hundreds) of times.
- Micro-rewards are constant: messages, notifications, updates, content.

These are your employees, your managers, and your future leaders.

When someone steps into an elevator and everyone is staring at their phones, that's not random. It's a habit loop. When you nearly bump into someone walking down the street because they're looking at their screen, that's a habit loop.

Now imagine you are competing with that for attention.

If your internal communications are dry, slow, and one-way while external platforms are immediate, interactive, and rewarding, where will your employees' best attention go?

Exactly.

The answer isn't to lecture people about "kids these days" or to reminisce about how hard you worked in your early career. No one cares about our mythical uphill trudges through snow and broken glass.

What people do care about is:

- whether they are informed
- whether their contributions matter
- whether participation is rewarded
- whether engagement feels meaningful

If you want to build powerful habits around internal platforms, you must ask:

- How do we make our internal tools worth returning to?
- What are the internal equivalents of "likes," "comments," and "shares"?
- How do we reward engagement in ways that build culture, not addiction?

That doesn't mean copying social media gimmicks. It means designing:

- recognition systems that are visible and fair
- feedback loops that are fast and useful
- communication channels that feel alive, not static

In other words, you are now in the business of designing **healthy dopamine loops** at work.

If your employees don't get rewarding experiences from you, they'll chase them elsewhere—often during the workday. That is not a character flaw. It's how our brains work.

Leaders who understand this will:

- treat attention as a strategic asset
- design internal experiences that are informative, interactive, and rewarding
- create a culture where participating in the company's story feels better than doomscrolling

Leaders who ignore it will keep sending long emails that go unread, then complain that "People just aren't engaged like they used to be."

Meanwhile, everyone's dopamine system is quietly voting with its feet.

9

Why Ingrained Habits Can Hinder AI's Ability to Scale Knowledge

AI's value comes from its ability to learn, adapt, and scale knowledge instantly. Human organizations, however, are built on routines designed for safety and predictability. Our ingrained habits—committees, pilots, meetings, escalation chains—are now one of the biggest constraints on AI-enabled progress.

The Real Bottleneck Isn't Technology—It's Us

If we're honest, humans are not built for rapid innovation. We debate. We escalate. We form committees. We run pilots over quarters or years. We argue about metrics. And eventually—after friction and trial and error—we move things forward.

This system worked for centuries.
It just doesn't work well tomorrow.

AI, however, advances without needing our rituals. Large-scale models scrape, synthesize, reason, and iterate without a task force or three-stage approval process. They improve every day whether humans are ready or not.

This gap—between how machines learn and how humans behave—is becoming the defining leadership challenge of the decade. It's our job to embrace this chance to disrupt bureaucracy.

The Rise of Agentic AI Systems

Stanford describes the next era of AI as one centered on **agentic systems**—software that doesn't just answer questions, but observes, plans, acts, and learns. They execute what humans typically slow down:

- logistics
- search and retrieval
- pattern detection
- coordination
- summarization
- experimentation
- progress monitoring

The breakthrough isn't that AI replaces people. It's that AI skips steps people invented for a slower era.

Humans still define direction and values.
AI handles the friction.

But that's only possible if leaders eliminate ingrained habits that slow adoption.

Where Habits Create Drag: Three Examples

1. Climate Innovation: Scaling What Works Faster Than Committees Can Decide

Climate initiatives often begin with vision, urgency—and then endless meetings. AI flips the script.

Instead of debating the future, an AI system could:

- analyze the top twenty-five climate interventions that already worked
- identify why they worked
- find regions with similar attributes
- model projected impact
- highlight where failure is more likely

Before a government or nonprofit writes a billion-dollar grant, AI could show where the idea is feasible, where it isn't, and why.

Humans: conceptualize ⇨ debate ⇨ pilot ⇨ scale

AI: analyze ⇨ match ⇨ scale

This is habit-breaking leadership.

2. Medicine: Scaling Expertise Beyond Physical Limits

We admire physicians for their dedication and intelligence, yet the system demands more of them than is humanly possible. One oncologist may treat 2000+ patients in a year. And while they all read the same journals, the real knowledge—the pattern recognition gained from treating individuals—remains trapped in local silos.

AI can change that.

Imagine if:

- a cardiologist in Jakarta could instantly access insights from ten thousand peers
- an oncologist could treat five more patients daily because AI preprocessed the case
- new physicians could get world-class second opinions instantly

AI becomes a **knowledge amplifier**, democratizing expertise rather than replacing it, and in the case of healthcare, expanding access.

3. Global Organizations: The Return of Focus

Executives today drown in:

- white papers
- emails
- notifications
- decks
- meetings
- new tools
- endless updates

In contrast, AI can:

- summarize every message
- consolidate information streams
- pre-read documents
- generate briefs
- synthesize project data
- surface risks
- explain options
- translate content into any format or medium

You wake up, pour your coffee, and your AI assistant proactively delivers exactly what matters.

Habit-breaking leadership means redesigning workflows around this new reality rather than forcing AI to fit existing routines.

Achieving AI's Potential Requires Human Change First

The ideas are not mysterious.

The technology is advancing fast.

The constraint is whether we reshape the habits that no longer serve us.

A knowledge revolution means:

- scaling climate solutions globally
- improving and expanding care for millions
- redirecting human time toward high-value thinking
- eliminating "busy work"
- making expertise universally accessible

All it requires is openness, a willingness to practice, and leaders who are ready to let go of the comfort of old routines.

AI is not waiting on technology.
AI is waiting on us.

10

The Importance of Muscle Memory

The real differentiator in the age of AI is the quality of our habits. Muscle memory will shape performance more in the future.

Why Practice Still Matters in the Age of AI

World-famous cellist Yo-Yo Ma reminds us that imagining practice forms the same neural pathways as physical practice. Elite performers know that consistency of preparation—not raw effort—is the secret to mastery.

Executives have the same challenge.

Over a forty-year career, a senior leader accumulates more than one hundred thousand hours of work. We read books, attend workshops, hire consultants—all in pursuit of higher productivity.

But as AI begins to remove the friction of information gathering, summarization, and execution, a quiet threat emerges:

If AI makes things too easy,
we may stop learning deeply.

AI will:

- find the right answer
- assemble the data

- format the brief
- analyze the trends

But if leaders don't practice the right mental motions, they lose the ability to judge quality, challenge assumptions, or strengthen intuition.

The Muscle Memory We Must Strengthen Now

1. Prepare Before Asking AI

Don't outsource thinking.
AI performs best when leaders frame the task clearly.

2. Understand the Platform

Most executives use 10–20 percent of what AI tools can do.
Explore the full capabilities.

3. Learn from Peers

Ask colleagues how they use AI.
Steal what works.

4. Reflect on Output Quality

What's missing?
What's unverified?
What's logically weak?

5. Never Assume AI Is Smarter Than You

AI is fast.
But humans decide what matters, what's ethical,
what's strategic.

The Leadership Imperative

Anyone who played a sport or practiced an instrument knows this truth:

Quality practice creates excellence over time.

In the AI era, leaders must practice:

- asking better questions
- framing problems
- checking assumptions
- reading nuance
- directing workflows

This is the "muscle memory" that will allow humans to scale alongside AI.

Yo-Yo Ma still practices three to six hours a day. The takeaway is simple: So should we—just in a different way.

PART THREE

Questions, Context, and Decision Quality

11

The Hidden Power of Questions

Future leaders will distinguish themselves by the precision and depth of the questions they ask. Questions, not answers, will become a fundamental unit of progress and a new executive superpower.

Questions Reduce Uncertainty—and AI Is Built on That Principle

Children ask hundreds of questions a day. Over time, adults unlearn that instinct—yet uncertainty remains the central obstacle in every organization.

Claude Shannon, the father of information theory, showed that communication quality depends on how effectively we reduce noise.

Questions are how humans reduce noise.

In AI, decision trees, Bayesian models, and machine learning all rely on progressively reducing uncertainty.

In business, leaders must do the same.

Why Questions Matter More in Organizations Today

We want to ask questions that:

- surface hidden information
- frame problems
- reveal assumptions
- align teams
- strengthen workflow
- activate innovation
- build trust

But corporate cultures often discourage them, mistaking speed for clarity and output for quality. We get right to "doing" without asking "why."

Two Examples of How Questions Drive Major Decisions

1. A $300 Million SAP Transformation

Twelve workstreams.

A hundred tasks per workstream.

Twelve hundred decision points.

If we identified the top five questions for each decision point and then shared this with all members of the project team, we would:

- move faster
- reduce rework

- avoid blind spots
- scale expertise
- eliminate dependence on tribal knowledge

AI can instantly surface the right questions at the right moment, based on our workload.

It can track what is working (or not) and share this in real time with other team members addressing the same topic (e.g., same questions and answers for each country) and learn from what has worked in peer companies.

Our job is to identify the right questions for each decision point.

2. Developing a New Drug

Let's assume there are 120 key decision points from preclinical to the Phase I clinical trial. At each step, we can pose the right questions at the right time related to:

- risk
- cost
- feasibility
- alternatives
- scientific rationale

AI can learn from subject matter experts, attach optimal questions to each time point, and store them to make us smarter with each trial.

This is a new form of organizational intelligence. Real-time improvements and a long-term shift in how we learn.

Why Our Questioning Muscles Are Weak

Keyword searches (Google-era habits) have atrophied our questioning ability. We just toss out words and get search results.

Generative AI is actually a mirror of our own intelligence. It requires:

- context
- specifics
- constraints
- goals
- reasoning

The better the question, the better the outcome.

We must retrain ourselves to become experts at understanding what we will ask and why.

Conclusion: Curious Companies Win

Curious companies will:

- finish SAP transformations early
- design better clinical trials
- accelerate discovery
- reduce risk
- teach teams faster
- elevate organizational intelligence

The question isn't whether we should do this. It's when we start. It is a market edge waiting to happen.

12

Questions Drive the Quality of Content

Questioning is a strategic competency.
It elevates the quality, accuracy, and usefulness of future work.
Not speed. Not money. Questions.

The Diagnostic Every Leader Should Try

Ask a leader if they're good at asking questions. They'll almost always say yes.

But ask them:

> ***"Do you know how to ask AI a question in different ways to get a more precise, strategic, or actionable answer?"***

Suddenly the room becomes quiet.

Great leaders don't give orders; they ask better questions. Leaders who are used to regularly asking questions of their team will be naturals for AI. Leaders who lead by command and control will find it more frustrating.

Five Models for Executive Questioning

1. Framing Model

Speak like you're briefing your top customer.

"Act as a lead oncologist and tell me the biggest unmet needs in cancer treatment today."

2. Logic Model

Lay out a structured request.

"Explain the sequence to launch a brand globally in eighteen months."

3. Comparative Model

Ask for tradeoffs.

"Compare $5M in advertising vs. a mixed media approach targeting one hundred thousand potential truck buyers."

4. Scenario Model

Test assumptions before a meeting.

"What market share shifts occur when a new competitor enters the men's deodorant category?"

5. Educational Model

Vary explanation by level.

"Explain small nuclear reactors like I'm a beginner, an intermediate learner, and an expert."

Practical Guidance

Assess your team's questioning patterns.

Teach them the five models.

Model it yourself.

Questions are a force multiplier.

13

Context Drives Content

AI is extraordinary at pattern recognition and historical analysis. But relevance, direction, and context remain fundamentally human responsibilities.
The question revolves around how leaders create the conditions for AI to produce meaningful work.

Why Context Is the Missing Ingredient

Imagine your enterprise employs ten thousand people—and ten thousand AI agents.

If those agents were a salesforce, you'd hold a national meeting to provide direction. But with AI agents, leaders often provide no direction at all.

Context is what differentiates relevance from noise.

What Leaders Must Provide

Leaders must shape:

- what questions agents should ask
- which data matters
- why a decision is important
- how tradeoffs should be weighed

- what the objective is
- which constraints apply

AI is not a mind reader.
It is a multiplier of whatever direction it receives.

It is our job to outline the responsibilities of each agent we employ. If we think of it this way, we actually have more control over the outcome than we may realize. And more responsibility to get it right.

The Leadership Imperative

We can't match AI on reach, but we can surpass it on relevance.

The quality of AI output is a reflection of the clarity of our intent.

14

The Shift from Attention Spans to Protecting Attention

It is convenient to complain about shrinking attention spans. But that's not the real issue. Leaders must redesign workflows so knowledge workers can do the deep, strategic, creative work they're hired for.

Attention Isn't the Problem—Environment Is

We scroll.

We swipe.

We skim.

We multitask.

But the goal of leadership isn't to improve attention spans. It's to **protect** attention so deep work is possible.

Knowledge workers are hired for their judgment, creativity, analysis, strategy, and nuance.

Yet we bury them in emails, chats, notifications, meetings, and interruptions.

This is a human failure.

What Deep Attention Requires

Deep work activates the prefrontal cortex, strengthens memory, improves problem solving, increases creativity, and enables strategic thinking.

Interruptions break this flow.

Research shows it takes approximately **twenty-three minutes** to re-enter deep focus after an interruption.

Now imagine an average day in your company. As leaders, are we unlocking the power of knowledge workers, or are we masters at interrupting their flow?

How Leaders Protect Attention

It's simple:

- reduce meetings
- streamline channels
- use AI to summarize inboxes
- empower AI to handle low-level work
- measure output, not hours
- encourage morning deep-work blocks
- remove friction
- enforce focus norms

Protecting attention is key to protecting talent and retaining knowledge workers.

Human-Only Capabilities Worth Protecting

Yes, AI will continue to learn from us, but if we are factoring in all the knowledge and nuances of a company, humans should retain an edge in the following areas:

- prioritization
- question formulation
- creative thinking
- reading nuance/reading a room
- judgment
- risk assessment
- values alignment

These human capabilities define organizational value and often help us create our market advantage.

AI elevates them when leaders protect the time required to use them.

Something I say often is that when we use AI effectively, it becomes a mirror of our intelligence.

15

How to Not Swallow the Ocean of AI

AI is advancing so quickly that leaders often feel an impulse to chase every headline, experiment, or new model. It is important for us to learn how to evaluate innovation without getting swept into hype, and how to build a repeatable process that keeps organizations focused, future-ready, and grounded.

Why AI Creates FOMO for Leaders

Every day brings a new AI breakthrough, platform, or concept. It's natural to feel pressure to respond immediately. But this instinct—to react quickly rather than think clearly—is the first trap.

When emerging technologies appear, society always imagines immediate transformation. We view tomorrow's tools through today's lens. That can lead to excitement or fear, but it rarely leads to clear thinking.

Headlines often focus on what AI will eliminate, not what it will create. Yet every meaningful technology wave—electricity, the internet, cloud computing—created far more jobs and industries than it destroyed.

The challenge is that humans live in the present, but technology evolves across years and decades. We skip steps because we want certainty now.

What Innovators Understand That Followers Do Not

Successful innovators look three to ten years ahead. They make decisions that shape markets long before the rest of the world catches up. They understand that:

- diffusion takes time
- adoption curves cannot be skipped
- most early ideas fail
- successful models require patience and iteration

The history of tech proves this. The first search engines appeared long before Google became dominant in 1997. Tim Berners-Lee created the World Wide Web in 1989, yet Facebook didn't appear until 2004. Between those moments were dozens of failed attempts—Orkut, Bebo, MySpace—each providing lessons the winners built upon.

Innovation is a decades-long journey disguised as an overnight success.

How Not to Swallow the Ocean

The real question isn't "What's new in AI?"

The best question is:

What advantage will this model create for customers, employees, or shareholders, and what will it take to scale it?

You don't need an MBA or PhD to evaluate this. You need a simple, disciplined process. Over decades of building models in and outside companies, I've found one approach that consistently works.

The NEXT Model: A Four-Step Process for Focused Innovation

This model keeps organizations grounded while moving quickly and intelligently. You don't need board-level approval. You just need to apply the logic of it.

1. N — Normal

How do we do this today?

Document the current approach in plain language so anyone can understand it. This baseline matters because transformation starts with seeing the present clearly. Most people know how to do something, but do not take the time to outline how they do it.

2. E — Experiment

What pilot will help us learn the truth?

Define:

- the questions you must answer
- metrics that illustrate customer value
- governance or risk considerations
- the timeframe
  (GenAI pilots: ~90 days; applied AI: longer)

Pilots are not experiments in hope. They are experiments in clarity, just like science, where we learn by iteration.

3. X — X-Factor

What makes the future state better?

Identify:

- what becomes easier
- what becomes faster
- what creates customer value
- what differentiates your organization

Speed only matters if customers benefit. If customers win, employees and shareholders usually do too.

4. T — Transform

Can we describe the new model clearly enough to scale it?

A scalable model should fit on:

- one page
- one slide

If the process is clear, simple, and compelling, it can be taught, adopted, and replicated globally. When you explain it, someone in that meeting can walk down the hall and repeat it accurately.

Simple scales. Complexity stalls.

Questions Every Leader Should Ask Daily

Because AI surfaces new possibilities constantly, leaders need a mental filter. Ask:

- Am I focused on what I can truly improve?
- Am I rushing and skipping steps, or am I following a disciplined model?
- Do my metrics reflect customer value and governance expectations?
- Does this idea deserve time, or am I reacting to hype?

If an idea feels chaotic, you may be trying to "swallow the ocean." Stop. Breathe. Focus on NEXT.

Innovation Is a Journey, Not a Reaction

Building and scaling meaningful ideas takes months or years, not days or weeks.

Leaders who stay disciplined will build the ships that outlast the storms.

Ten years from now, it will be obvious who applied clear models and who chased hype.

Be patient.

Stay focused.

Choose impact over excitement.

Learn from failure. Embrace it.

PART FOUR

Media, Creativity, and Communication in an AI World

16

The ABCDEs of AI and Its Impact on Media

Media behaviors evolve in predictable cycles, shaped by human attention, technology, and culture. The ABCDE Model offers a timeless method for crafting powerful narratives, now accelerated by AI. This chapter explains how AI is reshaping each component and what leaders must do to stay relevant.

Why Human Behavior Still Shapes Media

Technology doesn't determine how media evolves. People do.

Every time a new platform emerges—TikTok, Discord, Instagram—we collectively decide how to use it. We create norms, behaviors, and expectations that platforms then reinforce and amplify.

This predictable cycle makes storytelling more stable than most leaders realize.

That's why the **ABCDE Model*** remains the strongest framework for creating content that aligns with human behavior.

The ABCDE Model

A — Audience
B — Behavior
C — Content
D — Delivery
E — Evaluation

The most successful brands follow all five steps, in order.

You cannot:

- create great content without deeply understanding an audience
- build media plans without knowing where audiences gather
- expect positive behavior change without setting behavioral intent
- improve campaigns without evaluation rooted in the first four steps

AI accelerates all five, but the logic stays the same.

**The ABCDE Model was created by a group of teachers, inspired by how to build a powerful narrative for the United States of America.*

AI's Impact Through the ABCDE Lens

A — AUDIENCE:
The Marketplace Is Being Redefined Daily

Humans choose the channels that matter:

- TikTok became a global entertainment and search engine.
- Discord created new global audio-first communities.
- Instagram reinvented social commerce through influencers.

AI simply helps us understand these audiences and their potential faster, and at scale.

B — BEHAVIOR:
AI Is Reshaping How We Act Online

New behaviors emerge every year:

- real-time search everywhere
- conversational engagement with brands and apps
- shifting trust toward new influencers or information sources
- different purchase patterns
- evolving loyalty toward apps, not brands

AI reinforces behaviors by making them easier, faster, and more personalized.

C — CONTENT:
The Most Dynamic Part of the Ecosystem

If the marketing funnel collapses—and it is—then content must evolve with it.

Leaders must ask:

- How do we create trust instantly? And over time?
- Which words, colors, or images increase conversions?
- How should stories differ across channels like PI.ai vs. Instagram?
- How do we adapt content for platforms like Perplexity?

AI acts as:

- co-creator
- analyst
- strategist

- tester
- accelerator

The best creators today are part designer, part strategist, part behavioral scientist—and part futurist. The future Creative Director will become more of an Experience Director.

D — DELIVERY:
Media Plans Become Living Systems

With AI:

- Media plans will be updated more frequently.
- Channel sequencing will become more precise.
- Experiences will become deeper and more adaptive.
- Influencers will be advised by AI.
- In-app communities will matter as much as open social channels.

Delivery becomes dynamic instead of static.

E — EVALUATION:
Listening + Intelligence

Evaluation is shifting to:

- continuous listening
- behavioral intelligence
- agent-driven monitoring
- real-time supply chain tracking
- reseller and channel performance mapping

AI provides unprecedented visibility into how the marketplace behaves, changes, and responds.

The Role of the Communicator Is Evolving

The communicator of the future is part:

- detective
- forecaster
- behavioral economist
- community analyst
- content strategist

And they'll use AI to see around corners.

The ABCDE Model keeps the work grounded and ensures leaders never lose sight of the fundamentals.

What a ride the next decade will be.

17

A Lesson from Comedy (Reframing Issues)

Reframing is one of the most powerful tools leaders have. It helps teams see issues differently, break entrenched mental models, and unlock new solutions.

Comedic framing—the act of connecting two unrelated ideas—to public, social, and corporate challenges provides perspective on how we can reframe societal issues.

A Lesson from Comedy: Reframing Issues

John Cleese once said comedy works when you connect two completely different frameworks. The laugh occurs because the brain forms a new meaning at the collision point.

This is also how powerful reframing works.

We often lecture, over-explain, or drown audiences in complexity. Instead of creating clarity, we create distance. But when you pair two ideas that normally don't belong together, insight or action often emerges.

Example 1: Disinformation ⇨ Digital Pollution

"Disinformation is bad" is not compelling. It's vague and easy to ignore. But compare it to pollution.

We passed the Clean Air Act in 1970 because dangerous pollutants threatened our physical health.

Today, mis- and disinformation threaten our *cognitive* health. Our neurons can be polluted just like our rivers and air.

Once you frame disinformation as digital pollution, the problem becomes actionable. It becomes urgent. It becomes a public health issue.

Example 2: Cobalt ⇨ The Human Cost of EVs

EV batteries require cobalt. Much of that cobalt originates from artisanal mines that rely on child labor.

We love EVs. We love children.

Reframing the question—Does your EV use cobalt mined by a child?—creates a compelling moral lens that transcends statistics.

Example 3: Oncology ⇨ AI as a New Partner

There are ~19 million new cancer diagnoses each year and roughly 500,000–600,000 oncologists globally. Even if perfectly distributed (they are not), that's five thousand active patients per oncologist.

AI becomes an essential partner—not a replacement—simply because the math demands it.

Why Reframing Works

A framework is a set of assumptions. When you take the logic from one setting and apply it to another, you disrupt habitual thinking.

Comedy uses this to spark laughter. Leaders can use it to spark insight. AI makes it possible to figure this out, even if we aren't professional comedians.

18

The Shift from Data Lakes to Data Puddles

Most leaders misunderstand where AI's true advantages come from. The biggest breakthroughs won't come from massive public data lakes, but from smaller, curated, proprietary "data puddles."

Why Public Data Lakes Will Not Make You a Leader

Generative AI platforms pull material from the world's largest data lake—the internet. Everyone has access to the same data. That means you will have the same insights from the same content and draw similar conclusions to your competitors.

Yikes.

Leaders don't win by doing what everyone else is doing. They win by investing in proprietary advantages.

The Data Puddle Strategy

A **data puddle** is a curated dataset that:

- you select
- you quality control
- you de-noise
- you enrich
- you own

These puddles become your organization's unique data brain.

Take the best of open-source data, add it to your proprietary data, and give clear direction on what you desire, and you are building the ingredients for a strategic puddle.

Once AI has high-quality, differentiated data, it can produce insights no competitor can match.

Example: Metastatic Breast Cancer

Generative AI can pull from tens of thousands of clinical studies, but only ~10 percent ever succeed. Raw volume creates noise, not clarity.

A high-value **data pond** is formed by combining puddles such as:

1. Clinical trials (focused on the specific type of cancer, stage, and outcome)
2. Investigators (which ones were associated with successful trials)
3. Publications (who published, where was data presented)
4. Research centers (which centers shaped the market)
5. Approved treatments (how has FDA review and related guidance evolved for each treatment)
6. Real-world patient conversations and search patterns (what are people missing or appreciating)

Going back ten years gives longitudinal insight aligned to typical patent cycles. Precision will exist in how you instruct your agents.

This approach mirrors designing training sets for advanced machine learning.

Why Curated Data Wins

Exceptional algorithms share common traits:

- clear timeframes
- precise keyword profiles
- carefully weighted relevance
- curated media sources
- defined inclusion/exclusion criteria
- narrow or wide topic controls

With this structure:

- Generative AI becomes more accurate.
- Insights align with real-world evidence.
- Noise disappears.
- Content becomes more trustworthy.
- Science and communications accelerate.

You're not using AI to scrape the world—you're teaching AI your world. And it is helping you improve clinical trial design, select the best advisory board, understand your future sales strategy, and more.

Catching Up to AI's Potential

Generative AI is powerful, but it cannot advance science or communications strategy until leaders provide direction and differentiated data.

The technology is ready.
It's humans who need to catch up.

Leadership Lesson

The best models are simple.

Simple scales.
Complexity stalls.

Data puddles are simplicity with precision, and precision is where breakthroughs occur.

19

The New Résumé

Traditional résumés hide the skills that matter most in an AI-driven world. The next generation brings capabilities that don't show up on paper, but will redefine your organization.

Why the Résumé Is Failing Leaders

For centuries, résumés emphasized:

- where someone went to school
- where they worked and what roles they held

And that was enough—until now.

Today's rising generation has developed a skill set that is invisible in traditional formats. Many began building digital fluency at age ten. By the time they reach twenty-two, they are:

- multimodal learners
- creators
- editors
- coders
- analysts
- community members
- small-scale entrepreneurs
- fluent AI collaborators

None of this appears on a traditional résumé.

A New Kind of Talent

Young professionals today:

- learn from text, video, audio, apps, and games simultaneously
- produce and edit content with ease
- understand metrics instinctively
- collaborate across borders and time zones
- operate inside digital communities
- use AI tools as natural extensions of their workflow
- build and maintain personal brands
- develop real skills in gaming, streaming, coding, crypto, and ecommerce
- iterate based on engagement data, not managerial instruction

On day one, many could **teach** the organization how to work in this new environment. Of course, you can teach them on how the world works and how to understand your business. We need each other to unlock our collective potential.

Rethinking the Hiring Process

Consider two candidates for a creative solutions role:

- Candidate A: high GPA, strong school, traditional résumé
- Candidate B: average GPA, modest school, but ten years of Adobe experience, a website, and fluency in AI-assisted content creation

The better choice becomes obvious.

The résumé was misleading.
The skills that matter were hidden.

What Leaders Must Evaluate Instead

Ask:

- How does this candidate present ideas?
- What AI tools do they use, and what can they demonstrate?
- How do they partner with AI to expand capacity?
- What does their portfolio show?
- How do they think about personal branding?
- How do they learn new platforms?
- Can they explain their workflow?

This is talent scouting, not credential scanning.

Skills Learned from Ages 10–22 (That Résumés Miss)

- editing shorts (TikTok, YouTube)
- digital art and design tools
- troubleshooting IT
- systems thinking through gaming
- running Discord or online communities
- coding basics
- co-creating with ChatGPT and other platforms

- video production
- AI-driven design
- using AI tutors to accelerate learning

These are not hobbies.

They are workforce-ready capabilities.

The Leadership Imperative

Leaders must update how they evaluate future talent. The résumé must evolve to reflect capability, creativity, and digital fluency.

The next generation isn't less prepared.

They're more prepared than we realize, just in different ways.

20

Why Margin Improvement Requires Your Attention

AI isn't just accelerating productivity—it's exposing the hidden inefficiencies that leaders have tolerated for decades. The organizations that win in the AI era will be those that redefine work, redefine metrics, and redefine how time is valued.

We can rethink margin through the lens of workflow, incentives, and human behavior.

We Are Wired to Prioritize Effort Over Progress

For most of human history, survival depended on immediate action—find food, avoid danger, secure shelter. Our brains evolved to reward activity, not strategy. This is why:

- We equate effort with value.
- We feel guilty when tasks are completed "too fast."
- We fear being perceived as lazy.
- Dopamine reinforces task completion, not progress.
- We optimize for motion, not outcomes.

AI does the opposite.

It cares only about the goal—never the optics.

This creates a collision between **human habit** and **machine logic**, and that collision is already reshaping margins.

The Pressure Test for Today's Leader

Imagine leading a services firm that sells a "global blueprint" analysis for $37,500. Historically, it required 150 hours of billable time.

A rising star on your team uses AI—Perplexity, ChatGPT, and other tools—to produce superior output in 60 hours, complete with:

- a comprehensive analysis
- visuals
- a summary
- a narrated presentation deck

Mathematically:

- 60 hours × $250/hr = $15,000
- Standard 20% margin = $3,000
- Early completion "unlocks" $22,500 in additional profit

Total profit: $25,500 on a $35,000 project

This should be a win.

But it rarely happens.

Why?

Remember what I shared earlier in the book: Because the employee fears appearing unproductive. So they schedule meetings, involve more people, and stretch the work to fill the traditional timeline. The finance director doesn't adjust how they measure time and value.

Result?

The team overspends—$45,000 of internal cost on a $35,000 assignment.

AI didn't fail.

Our managerial habits did.

AI Reveals What Leaders Never Measured

AI shines a spotlight on the inefficiencies we've learned to ignore:

- unnecessary meetings
- inflated team sizes
- over-collaboration
- hours-based billing models that reward slowness
- workflows built around optics, not outcomes

To correct this, leaders must do what AI cannot—create new standards, new incentives, and new norms.

The Leadership Playbook for Margin Reinvention

1. Establish New Time Standards

Define the **expected AI-enabled timeline** for each project, not the legacy one.

The new definition of 100 percent efficiency is:

Meeting (or beating) the new AI-powered metric.

Not spending the full budget.

2. Hold Teams Accountable

Managers must coach teams to meet the new standards, not let legacy behavior creep back in.

3. Redeploy, Don't Idle

Fast completion should create capacity—capacity that gets **reallocated**, not wasted.

This is how companies expand offerings and grow revenue, not shrink.

The Fork in the Road for CEOs

There are two strategic paths:

A) Cut costs, shrink headcount, enjoy short-term margins

This "efficiency first" strategy works briefly—until the market reprices and competitors catch up.

B) Use the freed time to expand services and reinvent the business

This is harder, but it is where durable value is created.

The winners will be those who **treat saved time as a growth asset**, not an accounting gain.

The Real Leadership Message

Margin improvement isn't about squeezing more out of people. It is about **teaching teams a new way to work** and rewarding them for it.

AI shows us what is possible.

Leaders must show the organization what comes next.

21

A New Cousin of ROI: Return on Information

DuPont's ROI model reshaped management in the industrial age. "Return on Information" will reshape management in the AI age. It will force us to measure not just capital efficiency but thinking efficiency.

AI changes the economics of decision-making, requiring leaders to redesign questions, data, and workflows around it.

From Gunpowder to ROI

In 1914:

- Edward Bernays was twenty-three.
- Charlie Chaplin was debuting on film.
- DuPont was shifting from gunpowder to a broader industrial business.

To manage that complexity, DuPont developed a new model:

Return on Investment (ROI)

"Are we getting enough value for what we put in?"

ROI answered:

- How do we compare performance across plants, regions, and units?
- Where are we creating value?

- Where are we destroying it?

It was so effective that it:

- shaped generations of leaders
- embedded habits about cost, return, and capital allocation
- became a default mental model

The Limits of Traditional ROI

ROI excels at:

- hard assets—plants, equipment, inventory
- financial returns—margins, capital efficiency

It struggles with:

- brand equity
- employee engagement
- customer experience
- organizational intelligence

The variables that drive long-term value have always been harder to quantify.

Then along came AI.

A New Cousin Emerges: Return on Information

AI allows us to:

- mirror our own thinking patterns at scale
- capture decision flows
- automate parts of analysis and preparation

That creates the conditions for a new metric:

Return on Information – ROI (ai)
Precision / Volume × Decisions / Outcomes

In other words:

- Are we pulling in **precise** information, not just more of it?
- Are we cutting down **volume** that doesn't help?
- Are we improving the **decisions** we make and the **outcomes** we achieve?

ROI (ai) sits alongside traditional ROI and gradually influences it.

What Return on Information Will Change

Handled well, ROI (ai) will:

1. Re-center us on better questions

- Each role, level, and function will have "canonical questions" at key decision moments.
- AI can prompt those questions in the flow of work.

2. Shift us from "just in case" to "just in time" information

- Instead of months of analysis "just in case," we:
  - Build focused data puddles for recurring decisions.
  - Let questions trigger instant, tailored answers.

3. Simplify decision architecture

- AI can support more decentralized decisions.

- Hierarchies will matter less than **access to the right information at the right moment**.

4. Expose hidden pockets of value

- When decisions are no longer constrained by org charts, we see where:
  - Past habits are blocking growth.
  - Local teams see opportunities faster than headquarters.

Practical Applications

Identifying the Right Questions

Imagine every employee gets a prompt when starting a major task:

"Here are the eight questions leaders in your role typically need to answer at this stage."

For a clinical trial lead in breast cancer, for example, AI could surface:

- trial design precedents
- regulatory requirements
- patient population nuances
- competitive landscape summaries

Now imagine this at the scale of 100,000 employees across all repeatable decisions.

Preparing for Decisions

Rather than:

- Creating a huge custom research effort each time,
- We pre-build targeted data puddles for recurring decision types.

Then:

- A well-phrased question
- Instantly pulls together relevant data, insights, and options.

The hard part is:

- Building the puddles,
- Letting go of the idea that every answer requires months of human-driven preparation.

Extending Our Brains

Instead of "listening platforms" that track everything, ROI (ai) favors **intelligence systems** that track what matters *to you*:

- One leader wants to know what 1,500 vendors are doing globally.
- Another wants to know if a risk is trending across seventy-five languages.

AI and agents can:

- learn these preferences
- filter accordingly
- surface only what helps decisions

Expanding Innovation

Geography and language used to block ideas.

- An employee in Brazil with a great idea for a global product may never share it.
- They might worry about language, format, or politics.

AI can:

- translate the idea
- put it into the required format
- route it to the right people

We remove internal friction and discover innovation that previously stayed hidden.

Why This Matters Now

The original ROI model:

- clarified how capital was working for us
- defined the industrial era's management style

Return on Information will:

- clarify how **attention, insight, and decision-making** are working for us
- define how we manage in the AI era

Our scarcest resource is not capital. It's **time and cognitive bandwidth**.

As technology accelerates, our biggest obstacle isn't the tools—it's our habits.

ROI (ai) gives us a way to see and change those habits.

22

Agentic AI Ecosystems: The Biggest Disruptor of All Industries

It is easy to think of AI as a tool. But the most profound disruption will come when AI stops acting as a tool and starts acting as a ***worker****—a system of autonomous agents completing end-to-end tasks across domains.*

What Is an Agentic AI Ecosystem?

These are networks of AI agents that can:

- search
- compare
- negotiate
- purchase
- schedule
- monitor
- alert
- personalize
- optimize
- execute

...autonomously, with minimal human prompting.

They don't just answer questions; they complete **missions, while continually learning.**

Why This Is the Biggest Disruptor Across All Sectors

Because agentic ecosystems rewrite the logic of:

- search
- shopping
- customer service
- logistics
- media consumption
- healthcare navigation
- enterprise workflow
- sales
- marketing
- training
- operations

The companies that used to win by owning distribution, branding, or physical presence will lose ground to those who own **the AI interface**—the agent your customers trust.

The Power Shift: From Brand Loyalty to Agent Loyalty

In the traditional economy, customers follow brands.

In the agent economy, customers follow their personal agents, and those agents choose the brands.

The interface becomes the kingmaker.

This will be as big a shift as the move from desktop web ⇨ mobile web ⇨ apps ⇨ algorithms.

Except faster.

Leadership Implication

Companies must prepare now:

- build agent-ready product catalogs
- enable API-driven purchasing
- restructure support and commerce around machine-to-machine interaction
- create agent-safe content
- align compliance and governance frameworks

Those who wait will be disintermediated. Those who embrace will streamline commerce, find ways to build and scale brands that are not on our radar today, and uncover new market segments and opportunities.

23

How to Think Backwards to Imagine the Future: Agentic Commerce

Agentic commerce will feel "sudden" when it arrives, but we've actually been rehearsing for it for three decades. Your advantage comes from recognizing that what seems impossible today is often just the next logical step in a long chain of adoption.

A simple discipline involves thinking backwards from a future state to see how we're already preparing for it.

Touch, Confirm, Done: A Glimpse of Agentic Commerce

Imagine you're in ChatGPT looking at an image. You tap on a pair of shoes in the photo. An agent asks if you'd like to buy them. Behind the scenes, it:

- identifies the item and closest match to your preferences
- checks pricing and availability
- confirms your size based on past purchases
- shows you a quick summary and asks for approval
- places the order and schedules shipping
- tracks delivery and alerts you when it arrives

You only did two things:

1. **Touched the image**
2. **Approved the purchase**

It sounds extreme—until you notice how much groundwork has already been laid.

Thinking Backwards to See the Pattern

When something feels "too futuristic," I've learned to walk backwards through time and ask: *What had to be true for this to feel normal?*

Let's do that with agentic commerce.

1. 1994: "We'll Never Use Credit Cards Online."

In the early web:

- Using a credit card online sounded reckless.
- Fraud seemed inevitable.
- Most people assumed e-commerce would be a niche.

Then came Amazon and eBay in 1995.

Within a few years, using cards online became standard behavior. Fear gave way to convenience.

2. 2009: "We Need to Know It's Really You."

Next question: How do we know the person on this device is actually *you*?

- Two-step verification for Gmail in 2009 marked a turning point.
- Security shifted from "password only" to "prove it's you in a second way."

This ushered in a decade of identity verification innovation—codes, apps, tokens, and more.

3. 2011–2013: "We'll Never Trust the Cloud."

We used to say we'd never put sensitive information "in the cloud."

Then:

- Services like Google Drive and Dropbox normalized cloud storage.
- Many realized their data was safer in a professionally managed cloud than on a local hard drive.

We learned to trust that:

- Our files would be there.
- Access would be secure.
- The service would back things up better than we did.

4. 2017: "What If Someone Steals My Device?"

The next objection: What if someone steals my phone and pretends to be me?

- Facial recognition and biometric authentication answered that.
- Our face, fingerprint, or voice became part of the security stack.

Today, for many of us:

- our identity
- our payment methods
- our preferences

...all live safely behind **multiple layers of verification** we take for granted.

The Real Point: We've Been Training for This

When we look backward, we see a pattern:

- What once felt "crazy" becomes routine.
- Each "impossible" step is preceded by ten smaller steps that made it feel safe, normal, and convenient.

Agentic commerce—allowing agents to transact on our behalf—fits the same pattern:

- We already trust cloud infrastructure.
- We already store our payment data.
- We already rely on biometric verification.
- Merchants already operate inside this framework and can confirm who we are.

Both sides—buyer and seller—are operating on a shared security foundation.

Will there be hiccups? Of course. There always are.

The Discipline: Don't Just Look Forward, Walk Backward

Our biggest mistake as leaders is often hyper-focusing on the present:

- We list all the reasons something *cannot* happen.
- We overlook the decades of incremental progress that have quietly prepared us for it.

A better habit is:

1. **Imagine a future scenario** (like "tap to buy via agent").
2. **Walk backwards** and list the steps that would make it feel normal.
3. **Ask which of those steps already exist today.**

You'll often find we're much closer than your gut is telling you.

Technology is rarely magic. It's iterative, cumulative, and sneaky. When you learn to think backwards, you stop being surprised by the future and start designing for it.

Leadership Insight

The question is no longer:

"Will agentic commerce happen?"

It is:

*"How quickly will the market adopt it,
and how ready is my company?"*

The companies that prepare now will redefine global commerce. The ones who wait will be selling through someone else's agent.

I wonder what else we have been preparing to change over the last thirty years?

24

Embed the "Wow" in Workflow

Every major business era has introduced a new workflow discipline—Scientific Management, Total Quality Management, Six Sigma, Agile. AI will trigger the next one.

It will likely involve how we embed structure, speed, and "wow" moments directly into workflow.

A Century of Workflow Revolutions

- **1910s:** Scientific Management to scale manufacturing
- **1960s:** Total Quality Management and Just-in-Time to globalize
- **1980s–90s:** Six Sigma to handle complexity
- **2000s:** Agile to accelerate software

Today, AI is forcing the next leap—**workflow that is intelligent, adaptive, and partially automated.**

Why AI Demands a Structured Workflow Model

AI can:

- draft
- design

- analyze
- summarize
- route
- alert
- optimize
- personalize
- automate

But unless leaders redesign the workflow surrounding AI, its value stays trapped in isolated experiments.

What executives often get instead is:

- scattered pilots
- incompatible tools
- no integration
- inconsistent quality
- innovation theater

This is what slows transformation, not the tech.

The Components of the New Workflow Model

- **new habits** built through repetition
- **new metrics** focused on outcomes, not hours
- **new governance** to guide responsible use
- **new automation layers** to reduce human friction
- **new enablement** for teams to work with agents
- **new cross-functional alignment**

Workflow is the operating system for the AI-powered enterprise.

The Future Will Seem Obvious in Retrospect

Just as future generations laugh at our past fear of credit cards online, our grandchildren will shrug at how much friction existed in our workflows today.

They'll say:

"Why did you have to do that again?"

25

The Chief Workflow Officer

AI is reshaping every function, every process, and every job—simultaneously. No existing executive role has the mandate or span of control to manage this. Organizations need a new navigator: the Chief Workflow Officer.

Why This Role Is Emerging Now

Historical equivalents:

- Ancient era ⇨ The Scribe
- Industrial era ⇨ The Factory Manager
- Digital era ⇨ The CIO

The AI era requires someone who can coordinate:

- automation
- transformation
- behavior change
- governance
- cross-functional workflow
- training
- productivity design

...across the whole enterprise.

Enter the **Chief Workflow Officer (CWO).**

What the CWO Owns

- **Autonomous Workflow:** Which tasks should agents handle?
- **Knowledge Access:** How do we personalize intelligence for every role?
- **Prediction and Alerts:** How do we anticipate supply chain issues or valuation threats?
- **Boundaryless Operations:** How do we scale in 100+ countries with local nuance?
- **Automated Oversight:** How do we monitor policy, regulation, and risk?
- **Capability Building:** How do we elevate managers into AI-native leaders?

The CWO is not a headcount reducer.

They are a capability builder.

The Real Mandate

The CWO must:

- operate neutrally
- guide teams, not police them
- evolve habits over time
- teach leaders how to work with AI
- orchestrate change at a pace the organization can absorb

If the role becomes political, it fails.

If it becomes instructional and supportive, it transforms the company.

26

Building a Healthier AI-Driven Future

Healthcare has always evolved through innovation. AI is not an exception; it is an accelerant. Leaders who intentionally redesign workflow, training, and talent to work with AI—not around it—will shape a healthier, more equitable future.

From genomics and mRNA to robotic surgery and telemedicine, healthcare has repeatedly absorbed new technologies and translated them into better care. AI is the next major wave—only faster, broader, and more deeply embedded in daily work.

(From a keynote at the Healthcare Businesswomen's Association annual meeting 2025—illustrates application of concepts to a specific industry.)

1. The Chief Workflow Officer: A New Kind of Leader

Large organizations will increasingly need a **Chief Workflow Officer (CWO)**—a leader focused on:

- autonomous workflow management
- access to knowledge at the point of need
- alerts and predictions embedded in daily tools
- adapting to a boundaryless, hybrid organization
- automating oversight in highly regulated environments

But this is not only one person's job. Over time, **every leader and team member** will be expected to think in terms of workflow:

Where does AI remove friction, free up time, and improve care, and what do we do with the time we get back?

AI's potential is unlocked when our own habits and workflows are ready to match it.

2. Adding "Wow" to Workflow

AI is not just about automation and cost reduction. It is about **designing better experiences**:

- identifying and reinforcing new habits and skills
- using metrics to surface inefficiencies and opportunities earlier
- improving governance so AI use is safe, ethical, and compliant
- redesigning clinical trials with better site selection, eligibility criteria, and real-world inputs
- building dashboards that evolve with changing questions rather than staying frozen in time

Think of workflow as the "experience layer" for work. AI gives us the tools; leaders define the habits.

3. New Models to Reach Patients and Providers

AI is not something being "done to" healthcare. It is a set of tools and channels we can actively design.

Consider one example that sounds radical today but will feel normal tomorrow:

- By 2030, more than **four billion people** are expected to be gamers.
- Many of them will also be **patients, caregivers, and providers**.
- Organizations may appoint a **Chief Gaming Officer** responsible for reaching these audiences where they already spend time.

This role could oversee research, education, sales, and support inside virtual environments. If you're serious about access and engagement, you must be willing to follow your audience—into games, apps, and immersive experiences.

4. Training Will Be Continual, Not Episodic

The traditional model—annual trainings, compliance days, single workshops—will give way to continuous learning powered by AI:

- training that adapts to individual roles and current workload
- micro-lessons triggered by specific tasks or decisions
- AI coaches that help staff practice conversations, decisions, or procedures
- in-person programs that are now part of a broader, ongoing learning journey

AI makes it possible to treat every day as a learning day, not just the day of a scheduled training.

5. It Is Our Job to Guide AI

We cannot outsource our judgment to the technology.

Leaders will need to:

- develop a multi-year horizon for AI in their organization
- anticipate how policy, governance, ethics, and privacy will need to evolve
- decide where they want to lead vs. where they will follow
- establish clear guardrails and escalation paths

The mindset shift is important.

We won't merely react to AI; we will **shape how AI is used** in our organizations and across our industry.

6. Talent Assessment Will Evolve

The next generation of employees arrives with a very different skill set:

- editing short-form video and using digital art tools
- running fan groups on Discord or other communities
- producing AI-generated video, images, and narratives
- applying systems thinking through strategy games and simulations

Our job is not to dismiss these capabilities as "side hobbies," but to integrate them with:

- deep knowledge of science and regulation
- domain-specific expertise in disease areas, therapies, and care models

- an understanding of how healthcare systems function in practice

When we combine "new era" skills with existing expertise, we discover possibilities we don't yet have language for.

7. Forming New Habits (and Retiring Old Ones)

AI will change how time is spent:

- Some tasks will disappear.
- Others will be automated but supervised.
- New tasks—like prompt design, validation, escalation, and oversight—will emerge.

Leaders need to help teams:

- let go of low-value routines
- intentionally reallocate time to higher-value work (patient interaction, strategy, complex judgment)
- celebrate the creation of new habits, not just new tools

The question isn't just "What can AI do?" but **"What will we do with the human capacity it frees up?"**

8. Agentic Ecosystems Will Become Normal

We will increasingly rely on **agents**—AI systems that can act on our behalf within defined boundaries.

Examples include agents that:

- conduct first-pass reviews of documents or contracts

- benchmark pricing or terms for procurement
- draft responses or recommendations for human approval
- monitor regulatory or scientific updates and summarize key changes

The right way to think about agents is simple:

They are **additional team members**, waiting to be told what to do, with clear limits and accountability.

9. We Will Become Excellent Question-Askers

Getting value from AI will depend less on knowing code and more on knowing how to **ask better questions**:

- **framing questions** to explore context
  ("What are the three main paths we could take?")
- **logic-based questions** to test reasoning
  ("What assumptions are you making, and how would you challenge them?")
- **comparative questions** to evaluate options
  ("Compare option A vs. B on risk, cost, and speed.")

The organizations that win will build a culture where **curiosity and structured questioning** are everyday habits.

10. What We Won't Do

Equally important is what we decide **not** to do.

We will not:

- let "a thousand flowers bloom" with uncoordinated AI pilots everywhere

- tolerate isolated tools that don't integrate or scale
- accept inconsistent quality, unclear ownership, or "experiments" with no path to production

That approach leads to:

- fragmented tech stacks
- confusion in the workforce
- difficulty proving value

Instead, we will:

- prioritize a few clear workflows
- design end-to-end experiences
- measure impact and intentionally scale what works

11. The Opportunity in Front of Us

Healthcare has a strong track record of eventually absorbing innovation—from new drugs to new devices to new care models.

Now we have the opportunity to:

- improve how we manage workflow
- expand access to treatment
- improve outcomes
- make a meaningful difference in the trajectory of individual lives

As always, it starts with practice:

One step, one workflow, one habit at a time.

AI will be there alongside us—if we decide how we want to use it.

27

A New Era of Media

AI is reshaping not just how news is delivered,
but who delivers it, how it's consumed, how it's personalized,
and how deeply it integrates into daily life.
We are witnessing the reinvention of media distribution.

Five Reinventions That Will Redefine News

1. The New Delivery Person: Influencers

The newspaper on the doorstep is gone.

The new "local anchor" might be an Instagram, Facebook, or TikTok creator that we already trust.

Influencers become the last-mile distributors of local news.

2. The New CNN

Just as CNN used satellite technology to globalize news, modern platforms will use:

- WhatsApp groups
- Telegram channels
- gaming networks
- niche digital communities

News will follow where people already gather.

3. Aligning News with Personal Need

AI will:

- deliver rapid headlines
- dive deeper on command
- translate content into any language
- build personalized feeds
- reshape how journalists present their work

A single piece of journalism becomes a video, pictograph, podcast, or interactive Q&A.

4. The Conversational Newsroom

Readers will converse with their news source:

"Explain this like I'm ten."

"Show me the opposing viewpoint."

"What changed since last week?"

The outlet will remember what the user cares about.

This creates loyalty far beyond subscriptions.

5. Continuous Experimentation

News outlets will experiment with:

- custom search engines
- AI-generated visual explainers
- personalized front pages
- interactive editions for children
- multimodal storytelling

The Ted Turners of the AI age will emerge where experimentation is encouraged.

28

The Future of the Ad

Advertising will be reshaped by AI's ability to sense context, personalize delivery, and co-create with consumers. Creativity will expand—not shrink—as AI becomes an amplifier of human imagination.

Why This Future Is Different

Humans create from memory, intuition, and emotion.

AI creates from enormous datasets, pattern recognition, and generative capability and responds to our insights from memory, intuition, and emotion.

Together, they form a creative partnership unlike anything before.

What the Future of Advertising Looks Like

1. Behavior-Triggered Ads

AI understands:

- our activity
- our mood
- our environment

Campaigns change dynamically based on context—cycling race vs. casual ride.

2. AI Inside the Experience

Ads may become:

- a sponsored word
- a dynamic visual
- a personalized audio cue
- an in-chat recommendation

The **form** becomes the differentiator.

3. Your Personal Brand Ambassador

Consumers will choose the representative they want from each brand—a digital persona that remembers their preferences and guides decisions.

4. Storylines That Move Across Devices

Ads continue seamlessly from:

- phone ⇨ car ⇨ game ⇨ laptop

The campaign becomes a journey, not an interruption.

5. New Metrics and Optimization

We will optimize based on:

- biometric cues
- behavioral patterns
- IoT triggers
- edge AI signals
- location-aware personalization

6. Regulatory Complexity as a New Battleground

Regions will diverge sharply in AI advertising regulations. Campaign platforms will need to manage compliance at the city or building level.

Creativity Expands, Not Shrinks

AI amplifies the creative leader's imagination.

The next era of advertising will be more personalized, more contextual, and more immersive than any era before.

David Ogilvy would be fascinated.

29

The Next Breakthrough in Communications: The Knowledge Stack

Your role is no longer just to communicate.
Your role is to design a knowledge system that enables your teams—and your AI agents—to think, act, and improve together. The next breakthrough will not be a single person.
*It will be an **AI agent that continuously improves itself**, paired with leaders who know how to guide it.*

The Big Shift: From the Tech Stack to the Knowledge Stack

Every IT team in the world operates a **tech stack**—a structured system that enables creation, distribution, security, and scale.

But communicators rarely know what sits inside it:

- MongoDB
- Express.js
- EC2
- Node runtimes
- APIs

We don't need to be engineers.

But we do need something analogous—a way to structure the flow of ideas, insights, and information across an organization.

That is the **Knowledge Stack.**

Why We Need a Knowledge Stack Now

AI agents are accelerating decision-making, answering questions, synthesizing information, and writing code to perform new tasks. But they cannot improve without us.

Agents:

- rely on our guidance
- rely on our data
- rely on our questions
- rely on our clarity
- rely on our ability to teach them from successes and failures

Their improvement curve depends entirely on our intentional design.

This is why communicators, strategists, analysts, and operators must design the Knowledge Stack, not engineers alone.

The Components of the Knowledge Stack

1. Data Sources

The strength of your agents depends on the quality of data they can access:

- open-source media
- large public datasets

- internal proprietary information
- real-time signals (social, news, sentiment, search, forums, operations)

Think of yourself as a coach.

Your agents are your athletes.

Their potential must be trained.

2. Data Puddles: The New Unit of Innovation

For the first time in history, you can unlock innovation in every country—instantly.

- 350+ languages can be translated.
- Insights from Korea, Spain, or Bolivia can appear instantly in your dashboard.
- Local innovations can be shared globally in real time.

Innovation is now democratized.

A local idea instantly becomes global intelligence.

3. Key Questions

Questions are the engine of intelligence.

To analyze a crisis, a market, or a story in real time, leaders need to design the exact questions that unlock:

- what happened
- why it matters
- who is impacted
- what's the downstream risk
- what sensitive topics are involved
- what actions are required

AI agents don't know what to ask.

We must teach them.

4. Repetitive Actions

Every communications team has hidden repetition:

- preparing national sales meetings
- prepping executives for interviews
- updating media lists
- drafting Q&As
- analyzing weekly trends
- tracking clinical outreach
- producing summaries

Once identified, these actions can be automated.

Automation frees talent to focus on strategy, not clerical work.

5. Language Alignment

Your Knowledge Stack must speak a consistent language.

That includes:

- formatting
- metadata
- taxonomies
- naming conventions
- document structures
- tagging
- prompts

A shared language ensures that insights can travel across markets, teams, and AI systems.

6. What to Keep

Not every insight is valuable.

Not every signal matters.

The Knowledge Stack requires rules defining:

- which insights are worth preserving
- which insights drove meaningful outcomes
- which patterns improve future decisions
- which failures should be encoded as warnings

This teaches agents how to improve *and* helps leaders refine their own strategic instincts.

7. Gap Identification

Agents are exceptionally good at highlighting what's missing:

- missing data
- missing stakeholders
- missing perspectives
- missing trends
- missing risk factors

This is how we evolve from **reactive** to **anticipatory**.

The Future: Every Communications Leader Will Have a Knowledge Stack

Five years from now, the Knowledge Stack will be:

- a core competency
- a board-level conversation
- a competitive differentiator

- a requirement for speed
- a requirement for accuracy
- a requirement for global coordination

The timeline depends on one factor: **our willingness to build it.** It's not rocket science, but it is the foundation for what comes next.

30

The Missing Ingredient of a Powerful Story: Visual Narrative

Your customers don't just read your story—they watch it. Your narrative must be equally strong in words and images. Nearly everything in today's world is consumed visually, yet most corporate narratives are still built as if text is the primary vehicle for memory.

The next evolution of storytelling requires visual and textual narratives working as one, not operating in separate lanes.

The Brain Prefers Images, Not Text

Humans are wired for visual learning:

- We process images dramatically faster than text.
- We consume billions of hours of video daily.
- Platforms like YouTube, TikTok, and Instagram dominate discovery.

Despite this, corporate narratives rely on:

- eighty-page decks
- dense positioning documents
- long memos

- press releases
- endless emails

Text is vital, but it is no longer sufficient.

The Five Components of a Visual Narrative

1. Anchor Image or Memory

Humans often remember:

- a photo
- a metaphor
- a moment
- an image

...far more than text.

What is the memory you want your audience to hold?

This becomes your narrative's "center of gravity."

2. How You Unlock Interest and Emotion

Visuals can create:

- excitement
- empathy
- curiosity
- urgency
- surprise
- alarm

The goal is not to decorate your story.

The goal is to **introduce emotion intentionally.**

3. Message Alignment Across Channels

Every channel has a personality:

- YouTube ⇨ depth
- TikTok ⇨ immediacy
- Instagram ⇨ aesthetics
- Facebook ⇨ community
- Pinterest ⇨ aspiration
- Forbes ⇨ authority

Your visual narrative must adapt to each channel while reinforcing the same core message.

4. Rationale and Role of Visuals

Visuals should not be noise.

They should:

- simplify
- accelerate understanding
- reinforce trust
- break up cognitive fatigue
- enhance retention

Visuals must have **purpose**, not just presence.

5. Urgency, Timing, and Comprehension

In moments of urgency, visual first is often mandatory:

- product recalls
- crisis management
- health communication

- weather events
- executive announcements

Because visuals are interpreted faster, they outperform text when speed matters.

Steps to Build Your Visual Narrative Muscle

- Audit your current visual presence across the internet.
- Study how competitors use visual content.
- Identify the top 15–20 types of visual assets available—and use more of them.
- Track which visuals improve engagement or comprehension.
- Test new rhythms of storytelling across different channels.

You will quickly see that your brand has far more untapped visual potential than you realized.

The Future Is Multimodal

Generative AI now allows:

- instant translation
- instant visualization
- instant video summaries
- instant image generation
- instant story conversion into audio, slides, or pictographs

If customers prefer visual learning—and technology makes it effortless—the question becomes:

Why wouldn't you lead with visuals?

31

Designing for Absorption: How to Build More Innovative Teams

Innovation isn't just about ideas; it's about absorption—how quickly and deeply teams can take in the right information, make sense of it, and act. In the AI era, your competitive edge will depend on designing environments that maximize absorption, not just activity.

AI can help build immersive environments where people want to learn, think, and share.

What Meta Teaches Us About Subtle Shifts

Meta (Facebook) didn't win because of one big feature. It won by layering small shifts that increased **time spent, attention, and engagement**:

- the news feed
- the Like button
- comments and tagging
- groups and events
- acquisitions like Instagram and WhatsApp

None of these were dramatic on their own. Together, they:

- made the experience more immersive
- gave users reasons to return daily
- turned the platform into a default place to "check in" on the world

Most of us don't aspire to billions of daily active users. But we *can* learn from this:

Immersion is built, not declared.

Inside organizations, the equivalent isn't a social network—it's the **environment in which teams learn, share, and decide**.

Absorption: The Missing Metric

Companies and governments invest billions in:

- R&D
- strategy projects
- new technology platforms

Yet very few ask:

How long does it actually take our teams to absorb what they need and turn it into action?

Telling people to "work harder and smarter" doesn't fix this. AI alone doesn't fix it either.

What we need is a clearer understanding of **absorption**—the ability of teams to deeply engage with, understand, and apply the right insights at the right time.

When absorption improves:

- Decisions get better and faster.
- Innovation cycles shorten.
- The same people and budgets produce more value.

Designing Immersive Environments for Learning

Our brains absorb best when:

- We're immersed.
- We feel safe to explore.
- We feel in control of the experience.

An internal, AI-supported environment can create those conditions, if we design it with a few principles.

1. Relevance and Speed of Learning

People learn differently:

- Some prefer video, some audio, and some bullet points.
- Some want summaries; others want raw detail.
- Some need it in their native language.

Agents can:

- personalize format and depth
- deliver research in minutes instead of weeks
- adapt to preferences over time

The more relevant and faster the experience, the higher the absorption.

2. Democratizing Information

Imagine:

- A scientist in Japan and one at NIH access the same curated, high-quality data.
- Each adds their interpretation.
- Insights cross borders in real time.

AI can help:

- normalize access
- translate across languages
- surface patterns across locations and teams

Democratized access raises the baseline intelligence of the entire organization.

3. Building Data Ponds for Advantage

Rather than trying to ingest everything, you:

- curate the most important sources on key topics
- build focused data ponds (or puddles) around them
- let individuals query privately and save the most interesting insights

Over time:

- AI learns what each person cares about.
- The system starts to anticipate what to show next.

This is how you build **intelligence**, not just "content."

4. Ease of Use

If the experience isn't easy, it won't be used.

- Mobile access matters.
- Low-friction logins matter.
- Clear interfaces matter.

Absorption collapses when the tools get in the way.

5. Desire to Share

Innovation scales when people can:

- capture an idea quickly
- share it with the right audience
- have it automatically translated and summarized

AI can:

- turn a rough idea into a structured note or mini brief
- translate it for a global audience
- highlight emerging patterns across many small ideas

Sharing becomes a natural byproduct of how people work, not an extra task.

6. Privacy to Protect

Some ideas need:

- IP protection
- regulatory safeguards
- restricted access

Your environment should make it easy to:

- mark something as confidential with a tap
- keep it in protected zones
- still connect it to broader insights where appropriate

From "More Content" to "More Insight"

The goal is not to drown people in dashboards and feeds.

The goal is to:

- filter out noise
- filter in what's relevant
- highlight patterns that matter
- encourage depth over distraction

This is where AI can shine—as a **signal curator**, not just a content generator.

32

What Is Your New Creative Mix?

The AI era changes the creative equation.
You can now build campaigns that scale globally,
adapt instantly, and learn continuously from your audience.

Why We Favor the Status Quo

Our brains crave predictability:

- predictable results
- predictable processes
- predictable approvals
- predictable budgets

But predictability is no longer a competitive advantage.

Agility is.

The New Creative Mix: Three Layers

1. Professional Content (Your Story's Source Code)

Your narrative is the **source code**—the canonical version of the story.

Every channel, creator, and adaptation stems from this foundation.

2. Creator Content (Interpretation and Emotion)

Creators re-tell your narrative in ways that:

- feel authentic
- match platform norms
- build trust
- activate communities

You don't lose control—you gain amplification.

3. AI-Powered Scale

AI enables:

- translation
- voice cloning
- versioning
- cultural adaptation
- rapid experimentation
- real-time optimization

This is what turns a local campaign into a global one.

Governance: Fix and Remix

The creative team continuously:

- tracks what works
- stops what doesn't within hours
- remixes what resonates
- amplifies what converts
- updates briefs based on real-time audience data

Campaigns get **stronger over time**, not weaker.

What AI Will and Won't Replace

AI will not replace:

- big ideas
- creative direction
- human insight
- community understanding
- storytelling instincts

AI will replace or accelerate:

- asset generation
- performance analytics
- global scaling
- influencer matching
- narrative optimization
- real-time dashboards
- versioning and personalization

The Creative Mix Becomes a Partnership

Your customers become co-creators

Your creators become your distribution.

AI becomes your multiplier.

When combined, this is the most powerful creative system we've ever had.

33

Reinvent Paid Media by Focusing on Who Delivers the Story

Influencers are not an add-on. They are now the primary delivery system for many audiences, and paid media strategies must reorganize around them. For decades, paid media optimized for impressions, reach, and frequency.

The rule of seven (seven exposures to drive action) was a convenient average, not a truth. But now, AI, analytics, and creator ecosystems are rewriting how paid media works.

The Big Shift: Delivery > Frequency

Traditional advertising assumed:

- audiences were passive
- brands had to repeat the message
- trust was built through exposure

Today:

- audiences self-identify by following creators
- creators hold trust and trust drives action
- action is the new metric (not exposure)

Creators *are* the marketing funnel.

Audience Architecture: Who Drives and Who Shapes a Market

In any market:

- **1%** drives (influencers and creators)
- **9%** shapes (engaged followers)
- **90%** adopts

AI helps you map:

- every creator
- every micro-segment
- every topic cluster
- every behavioral group

This enables precision at a level advertisers never had before.

A New Paid Media Model

A company targeting small businesses might find:

- thirty creators reaching fifty million people
- one hundred micro-segments aligned to one product
- potential revenue of $25M from one hundred thousand converted customers

AI then matches:

- the right creator
- to the right audience
- with the right message
- at the right time

Creators get compensated dynamically based on performance. Paid media becomes a **real-time experience team**, not a placement exercise.

The New Metrics

Old metric: exposures

New metric: **movement**

Success becomes:

- action taken
- conversion per micro-segment
- audience expansion
- creator effectiveness
- experience resonance

You join their ecosystem, not the other way around.

34

There Are No Barriers to Launching Global Content Except Us

AI removed the barriers of language, cost, time, and production. The only barrier left is our reluctance to use these tools boldly. For the first time in history, a brand can scale content globally in minutes, with native voice, cultural nuance, and high-quality production.

Leaders must redesign their content strategies to take advantage of this superpower.

Why This Moment Is Different

Platforms like ElevenLabs now allow:

- natural voice cloning
- accurate emotion replication
- synchronized lip movements
- instant translation
- rapid localization

This transforms how brands launch globally.

A New Launch Model

Imagine:

- one two-minute video
- instantly translated into twenty, fifty, or one hundred languages
- your exact voice, tone, and emotion
- local nuance preserved
- insights gathered in real time
- continuous iteration
- global brand with local soul

This was impossible until now.

The Barriers That Just Disappeared

- language
- production cost
- time
- access to local nuance
- distribution friction

What's left? Execution and imagination.

Global Content Is Now a Two-Way Conversation

You can now:

- launch
- listen

- adapt
- personalize
- co-create
- localize
- evolve

All in real time.

This changes the relationship between a brand and its audience forever.

35

How to Cut Through the Noise and Innovate

Innovation doesn't come from reacting to noise. It comes from identifying one enduring trend, building diverse teams, seeking simplicity, and believing that change is both possible and necessary. Every major technological disruption—from Napster to iPhone to generative AI—follows the same pattern.

Most leaders panic. A few remain calm, read the trend correctly, and build the future.

The Four-Step Playbook of True Innovators

1. Identify the One Trend That Matters

Napster revealed the trend:

Consumers want to curate their own music.

It is an irreversible truth.

Similarly, with AI:

Consumers want personalized, immediate, agentic intelligence.

2. Build an Intellectually Diverse Team

Breakthroughs happen when software engineers, communicators, strategists and executives collide in productive chaos.

Like electrons jumping orbit, the team appears discontinuous but is actually advancing together.

3. Recognize That the Best Solutions Look Simple

Spotify's idea was simple.

The iPhone was simple.

The best AI products will be simple.

Simple is not simplistic.

Simple is clarity earned through iteration.

4. Believe You Can Change How People Work or Play

Every transformative product:

- reshapes a habit
- rewires a behavior
- builds a new expectation

AI is doing this right now with:

- voice interfaces
- agentic recommendations
- multimodal learning
- personalized content
- predictive culture

The innovators of this era will be those who design **how people interact with knowledge.**

36

The New Marketing Funnel: An Early Look

The funnel is no longer linear—AI-powered agents now shape every stage of the decision cycle. AI collapses awareness, consideration, intent, and purchase into a dynamic, always-on loop. Brands must learn how to thrive inside the agent-driven ecosystem.

The AI-Era Decision Loop

1. Trigger ⇨ Intent

The customer sees a need or an agent detects it before the customer. Either works.

Brand requirement:

Be discoverable by agents. The more visible, the more an agent can help.

2. Agentic Filtering

Agents curate reviews, claims, comparisons, and expert sources to produce a shortlist.

Brand requirement:

Optimize for AI retrieval and trust.

3. Micro-Decision Moments

How do we build trust, align values and timing, and provide the right incentives?

Brand requirement:

We design and prepare for micro-moments, which are all part of a single conversion.

4. Personalized Action

Agents may execute the purchase, complete the delivery, and send us the receipt.

Brand requirement:

One-click leads to a complete transaction.

5. Feedback ⇨ Continuous Learning

Everything loops back to the agent, which will have a memory of our interaction.

Brand requirement:

Customer support, the next purchase, all part of an agent-based relationship.

37

Trust Is Not Decreasing — It Is Diffusing

The narrative that "trust is dying" is inaccurate.
Trust is being redistributed, reshaped, and personalized.
AI accelerates this shift by expanding the number of people, platforms, and micro-communities individuals can rely on.
How we build trust is changing.

Trust Is Diffusing, Not Declining

Headlines routinely say trust is falling—trust in Congress, scientists, CEOs, journalists. But the data obscures a far more important truth:

Trust is not decreasing. It is diffusing—flowing to new people, new communities, and new environments.

Humans are wired to trust those who are present in our daily lives. Psychologists call this **availability bias**. We tend to believe the people we hear from regularly, the people who simplify our world, the people who feel familiar.

Digital life has dramatically expanded those from whom we "hear."

And this is the source of the shift.

Why We Missed This Shift

We measure trust the way we measured it twenty years ago—by surveying faith in institutions, government, business leaders, and media.

But the "n"—the independent variable—has changed.

In a world of billions of digital voices, the trust pool is simply growing larger and more distributed.

Three structural changes explain why.

1. The Freedom to Follow

Six billion people can choose exactly whom they follow.

And when given the choice, humans overwhelmingly prefer:

- People ⇨ not companies
- Personalities ⇨ not institutions
- Creators ⇨ not brands

This is why:

- TikTok became a global force.
- Instagram creators outweigh corporate accounts.
- Governments or companies rarely appear in the list of top twenty-five influencers in any major country.

We built a new, decentralized media distribution system.

Executives who ignore it will misread how influence actually works.

2. Micro-Communities Become the New Social Fabric

Gaming platforms, Discord channels, private groups—people now spend hours in micro-communities defined by shared interest, not geography.

These groups form persistent digital companionships:

- friends in ten cities gaming together
- hobbyists sharing daily insights
- fandoms functioning as social identity groups

When people choose their own micro-communities, they choose their own trust centers.

Trust doesn't shrink—it reorganizes.

3. Encrypted Echo Chambers

Private messaging groups have become the new "living rooms" of the digital age. People want safety, familiarity, and a space free from public criticism.

These encrypted circles create small islands of high trust even as institutional trust fluctuates.

The result?

People trust individuals, micro-communities and closed groups more than formal institutions.

The Leadership Implication

Trust exists in abundance.

Just not always where we expect it.

Leaders must build **new models, new metrics, and new maps of influence**. The organizations that thrive will understand:

- where trust lives
- who has influence
- which micro-communities matter
- how trust flows across networks

Trust didn't die. It moved.

Your relevance depends on learning where.

PART FIVE

Security, Ethics, and Governance

38

The Other Half of Security: Battling Bad Actors

Cybersecurity goes beyond servers and firewalls.
The new threat surface includes your leaders' voices,
your customers' beliefs, and your reputation.
The modern security organization must integrate cybersecurity
and cognitive security into one unified discipline.

Bad actors are adopting AI at the same speed—
or faster—than legitimate enterprises.
Their advantage: no regulations, no ethics, no limits.

The Threat Is Expanding

The vocabulary hasn't changed:

- breach
- attack
- threat

But their impact has grown.

Bad actors now aim to:

- destabilize trust
- create market panic

- erode reputations
- mimic leaders
- manipulate public perception
- blackmail companies into silence

They may not attack your servers.

They may attack your reality.

Misinformation vs. Disinformation

Disinformation = intent to harm

Misinformation = intent to share, even if false

Both can damage companies.

But disinformation is becoming industrialized.

Examples of AI-Enabled Attacks

1. Replace Your Narrative

A pharma company launches a breakthrough cancer vaccine. Bad actors fabricate stories about:

- faulty technology
- bribed executives
- fatal clinical outcomes

All false, all repeated daily for months.

2. Spark Global Panic

A bad actor uses cloned voices of mid-level government officials to put out false information of financial risk. It shows up as:

- Brazil ⇨ an audio leak
- Spain ⇨ audio + video
- South Africa ⇨ coordinated "reports"

The false message:

Your bank is about to collapse.

Markets react. Governments issue statements. Trust erodes.

3. Extortion Evolves

Instead of hacking your systems, bad actors say:

"Pay us to stop the narrative attack."

This will become common.

What Leaders Must Prepare For

A modern digital security program must anticipate:

- deepfake detection and anomaly spotting
- protection of executive likeness and voice
- modeling future influence-based attacks
- early-warning systems using AI intelligence
- understanding adversary technology
- autonomous influence campaigns

Security and communications are merging.

The Basic Psychology Bad Actors Exploit

- **desensitization**—repeated exposure breeds belief
- **long-term conditioning**—emotional fatigue weakens skepticism
- **multiple sources**—falsehoods seem credible when echoed
- **groupthink**—communities normalize false narratives

Understanding the psychology is as important as the technology.

The Leadership Message

Bad actors do not play by rules.

They do not respect borders.

They do not slow down.

Organizations must:

- broaden the definition of security
- integrate communication, IT, and intelligence teams
- build cognitive and reputational defense capabilities
- anticipate—not react to—threats

This is the other half of security.

And it deserves our attention.

Bad actors use the same technology we use with one major advantage. They do not have rules, guidelines, ethics, or a conscience that gets in their way, which means we have to outthink their next moves to protect ourselves.

39

The Importance of Cognitive Security

Your employees' minds—not your servers—are the most vulnerable attack vector. Cognitive security is becoming a core pillar of organizational defense. AI supercharges the spread of mis/disinformation, deepfakes, and emotional manipulation.

Why Cognitive Security Matters

We can recover from a hack.

A changed belief is much harder to correct.

Adversaries target:

- perception
- attention
- emotion
- identity
- trust

This is the most effective, lowest-cost attack surface.

A "Slow Burn" Attack

Most influence campaigns begin quietly.

Signals are subtle.

Patterns emerge slowly.

By the time leadership notices:

- sales decline
- talent starts leaving
- markets lose confidence
- customers lose trust

The campaign may have been active for years.

Cognitive Vulnerabilities

Enduring patterns of biased thinking—such as negative beliefs, attentional biases, or rigid interpretations—that make a person more susceptible to stress and psychological distress.

Cognitive Load Theory

Too much information overloads working memory ⇨ poor judgment.

System 1 / System 2 Thinking (Kahneman and Tversky)

Disinformation forces people into:

- System 1 ⇨ fast, emotional, impulsive

instead of

- System 2 ⇨ slow, analytical

Social Identity Theory

People trust their group and reject outsiders.

Peripheral vs. Central Processing (Petty and Cacioppo)

Some respond to detailed explanations; others respond to trusted messengers.

Cognitive Dissonance (Festinger)

People hold beliefs or opinions that contradict, which leads to mental discomfort. As a result, they often avoid information that contradicts their beliefs.

Deepfakes = Cognitive Pollution

Even when people *know* something might be fake, exposure affects:

- confidence
- perception
- trust
- memory

This is why deepfakes are powerful—even when unbelievable.

Daily Cognitive Overload

Your employees receive:

- 500–1,500 daily digital exposures
- 10–30 percent contain inaccurate information
- hundreds of social interactions
- dozens of notifications

This constant noise makes them vulnerable.

The Leadership Response

Leaders must:

- teach employees cognitive security
- build early-signal detection models
- treat mis/disinformation as strategic threats
- train teams in psychological literacy
- monitor narrative environments
- build protective habits

Security is no longer just technical.

It is cognitive.

40

Truth Seeking and the Disinformation Alarm

Disinformation is no longer episodic—it is continuous, evolving, and increasingly invisible. The future requires new detection models, new forms of public education, and new collaboration across institutions. Most organizations focus on the visible surface of disinformation—content that goes viral.

But the most dangerous forms are below the surface: censorship, algorithmic manipulation, youth exposure, cross-border coordination, and AI-enabled influence.

The Iceberg Model of Disinformation

Above the surface = the posts we see.

Below the surface = the forces shaping what we *don't* see.

The real threat is hidden beneath the waterline.

Five Alarming Shifts Leaders Must Understand

1. Innovative Censorship

Censorship is no longer brute-force deletion. It includes:

- shadow banning
- account suppression

- algorithmic favoritism
- selective invisibility

We spend too much time analyzing what is said and too little time analyzing **what never reaches people at all.**

2. Gaming Is the New Family Room

Nearly four billion gamers treat platforms as:

- communication hubs
- content ecosystems
- commerce platforms
- social networks

Bad actors will use gaming the way nation-states use social media today.

3. Long-Term Cognitive Impact on Youth

Young brains are still forming.

Constant disinformation leaves:

- cognitive scars
- trust erosion
- worldview distortion
- emotional fatigue

Bad actors are playing the long game.

4. Fragmented Good Actors, Unified Bad Actors

Good actors:

- don't share knowledge
- operate independently
- use inconsistent strategies

Bad actors:

- coordinate
- optimize messaging
- share tactics
- amplify each other

Coordination is a competitive advantage.

5. Technology Is Advancing Faster Than Our Defenses

Bad actors now have:

- billions of synthetic identities
- instant voice and face cloning
- compressed audio masking
- real-time narrative manipulation

We are entering Disinformation 2.0.

The Path Forward

Leaders must:

- build R&D pipelines for adversary modeling
- create open-source "censor-check" capabilities
- design gaming-era interventions
- teach digital media fluency
- collaborate across borders
- assume attackers are always a step ahead

If we want truth-seeking, we must build alarm systems.

41

The Future of Section 230

Section 230 is entering a new era. Generative AI will force governments to reconsider whether platforms that produce content should receive the same protections as platforms that merely host it. Section 230 enabled the growth of the modern internet.

But AI changes the nature of content creation, responsibility, and risk.

Why Section 230 Matters

The core protection:

"No provider...shall be treated as the publisher of content from another provider."

This shield allowed:

- social platforms
- forums
- user-generated content
- digital communities

...to flourish.

Why AI Disrupts This Framework

Generative AI:

- creates original content
- adapts content
- personalizes content
- responds autonomously
- influences perception

Platforms are no longer **hosting** content—they are **enabling the creation of it**.

This raises new questions:

- If a model generates harmful content, who is responsible?
- If an AI system amplifies hate speech, is the platform liable?
- Should the protections of 1995 apply to the AI systems of 2025?

What is the right legislative solution?

The Policy Guardrails Taking Shape

Future legislation will likely:

- distinguish between hosting and generating
- emphasize transparency
- update liability protections
- support economic growth while reducing harm

- incentivize content provenance and watermarking
- hold platforms accountable for algorithmic amplification

Section 230 will not disappear.

But it is difficult for a 1995 law to effectively govern a 2026 technology.

New bipartisan legislation is only a matter of time.

42

Podcast Summary of Talk with Theologian Ted Vial: AI and Ethics

AI ethics is human ethics. The future requires frameworks that combine philosophy, psychology, governance, and technology. Ethics in the AI era is about human choice, human responsibility, and human judgment, not mathematical formulas.

Key Lessons from the Conversation

1. AI Ethics = Human Ethics

Humans—not machines—must:

- set standards
- define fairness
- interpret impact
- regulate use

Technology cannot solve moral disagreements.

2. Math Cannot Eliminate Bias

Algorithms reflect:

- data
- assumptions

- incentives
- cultural norms

Fairness is not a math problem—it is a societal negotiation.

3. Professions Require Standards

Doctors must be licensed.

Engineers must be trained.

Pilots must be certified.

What is the equivalent for people writing code that affects millions?

This is where ethical frameworks are heading.

4. Algorithms Are the New Editors

Algorithms decide:

- what we see
- what we don't
- what is amplified
- what is hidden

Editors used to be named individuals.

Now they are lines of code.

Transparency becomes a moral requirement.

5. Gaming Teaches Connection

Gaming is not just entertainment.

It is global, social, collaborative learning at scale.

Connection builds empathy.

Empathy builds trust.

6. Availability Bias Shapes Belief

People tend to believe:

- what they see repeatedly
- what familiar messengers say
- what their community reinforces

Ethics must account for human psychology.

7. Privacy Means Something Different Everywhere

In Europe, it means consent and data rights.

In much of the world, it means physical safety.

Ethics must be global.

Technology must respect cultural context.

Final Reflection

What seems new—consciousness, fairness, moral AI—has been debated for centuries.

Ethics is not a technical problem to solve.

It is a human responsibility to uphold.

43

Understanding the Foundation of Ethics Debates

Ethics debates are rarely "new." Whether you're talking about AI, climate policy, health equity, or content moderation, the same core questions repeat: freedom vs. control, fairness vs. efficiency, innovation vs. harm. If you understand these foundations, you can stay ahead of ethical issues instead of reacting to them one headline at a time. When new technologies like AI arrive, it can feel like we're facing completely new moral terrain. In reality, most debates track back to a familiar set of themes.

Think of this as your ***ethics lens****—a way to frame conversations, anticipate objections, and design better policies before you're in crisis mode.*

The Ten Pillars of Ethics Debates

Instead of getting blindsided issue by issue, it helps to recognize the **recurring patterns** that show up in almost every ethics conversation.

1. Autonomy and Individual Rights

- **Freedom vs. control:** How much should people control their own choices vs. society imposing guardrails (free speech, medical consent, data privacy)?

- **Informed consent:** Do people truly understand what they're agreeing to—whether it's a surgery, an AI terms of service, or participation in a study?
- **Personal responsibility:** When individuals make harmful choices, how accountable are they vs. the systems around them?

2. Justice, Fairness, and Equity

- **Distribution of resources:** Who gets access to wealth, healthcare, education, or new technologies?
- **Bias and discrimination:** Are we embedding or correcting inequalities across race, gender, income, or geography?
- **Global justice:** How do we balance the needs and burdens of rich vs. poor nations, especially for climate, health, and digital access?

3. Beneficence and Nonmaleficence

- **Do good, avoid harm:** The classic medical principle now applies everywhere—from environmental impacts to product safety to algorithm design.
- **Risk vs. reward:** When is it ethical to push ahead with innovation (AI, biotech, nuclear, geoengineering) if there is real potential for harm?
- **Negligence:** When accidents happen, who is morally—and legally—responsible?

4. Truth, Honesty, and Transparency

- **Deception and manipulation:** Marketing, politics, media, and now AI-generated content all test our commitment to truth.

- **Information integrity:** Misinformation, disinformation, and "data pollution" shape how entire populations see reality.
- **Trustworthiness:** Sustainable systems require transparency and explainability to earn social trust.

5. Power, Control, and Governance

- **Who decides?:** Many ethical issues boil down to who has the power to make and enforce rules.
- **Corporate vs. public interest:** Are we optimizing for profit, public good, or some mix of the two—and who decides the mix?
- **Accountability in leadership:** When leaders misuse power, how should they be judged, constrained, or removed?

6. Human Dignity and Respect

- **Personhood:** What counts as a "someone," not a "something"? This shows up in debates about AI, animals, embryos, and end-of-life care.
- **Privacy:** How do we respect personal boundaries in a surveillance-heavy, data-driven world?
- **Exploitation:** Where are people being treated as a means to an end—whether in sweatshops, the gig economy, or digital labor?

7. Environmental and Intergenerational Ethics

- **Sustainability:** Are we leaving the planet livable for future generations?
- **Climate justice:** Who bears the burden for damage already done and sacrifices needed going forward?

- **Stewardship:** What is our responsibility to ecosystems and non-human life?

8. Technology and AI Ethics (Baseline)

- **Algorithmic bias:** How do we ensure systems are fair across different populations?
- **Autonomy vs. automation:** When does replacing human judgment with AI cross a line?
- **Surveillance and privacy:** What trade-offs are acceptable between safety, convenience, and data collection?

9. Bioethics and Human Enhancement

- **Experimentation and consent:** Are patients and participants protected and well-informed?
- **Genetic engineering and cloning:** Just because we can, should we?
- **Life extension:** What are the implications if we dramatically alter life expectancy or enhancement access?
- **Equity:** Who benefits from advanced therapies and who is left out?

10. Meta-Ethics and Moral Relativism

- **Universal vs. cultural values:** Are there truly universal human rights, or is morality context-dependent?
- **Objective vs. subjective morality:** Is "good" something real, or something we collectively agree on?
- **Moral progress:** How do societies decide that past norms (slavery, discrimination, unsafe workplaces) are no longer acceptable?

II. How AI Stretches These Ethics Foundations

AI doesn't invent ethics from scratch, but it **tests the limits** of existing frameworks. Here are the key debate zones leaders will encounter.

1. Agency and Autonomy of AI

- **Moral status of AI:** If advanced systems appear conscious, do they deserve any moral consideration—rights, responsibilities, or protections?
- **AI decision-making:** Should autonomous cars, trading systems, or weapons make life-and-death decisions without a human in the loop?
- **Delegated agency:** If my AI negotiates, recommends, or even votes for me someday, where does my responsibility end and the system's begin?

2. Accountability and Liability

- Who is responsible when AI harms someone?
  - The developer?
  - The deploying company?
  - The end user?
  - The regulator?
- **Diffused responsibility:** AI supply chains blur accountability in ways we've never seen before.

3. Value Alignment and Control

- **Alignment problem:** How do we ensure AI systems pursue human values, not just efficiency or narrowly defined objectives?

- **Mis-specified goals:** Even small mistakes in what we ask for can lead to catastrophic behavior at scale.
- **Control problem:** Can humans maintain meaningful control over increasingly capable, autonomous systems?

4. **AI Bias and Fairness at Scale**

- **Bias amplification:** Models can replicate and amplify human biases in hiring, policing, credit, and healthcare.
- **Synthetic bias:** AI may generate new forms of unfairness not traceable to a single human decision.
- **Invisible discrimination:** Harm can become subtle, statistical, and hard to detect without careful monitoring.

5. **Human Identity and Agency**

- **Skill erosion:** If AI tutors, copilots, and diagnostic tools do the thinking, do human skills atrophy?
- **Authenticity:** When art, writing, and even "friends" can be AI-generated, what counts as "real"?
- **AI companions:** Emotional bonds with machines raise questions about manipulation, loneliness, and dependency.

6. **Information Integrity and Deepfakes**

- **Epistemic crisis:** If any photo, video, or voice can be faked, how do we know what to trust?
- **Democratic risk:** AI-generated propaganda and synthetic media can destabilize elections and governance.
- **Cognitive security:** We will need new approaches to protect human attention and reasoning from hyper-targeted, AI-optimized manipulation.

7. **Labor, Economy, and Inequality**

- **Job displacement:** This time, automation affects knowledge workers, not just manual labor.
- **Power concentration:** AI capabilities and data are clustered in a small number of companies and countries, widening global gaps.
- **New economic models:** Labor, creativity, and data may be tokenized or compensated in new ways.

8. **Existential Risk and Long-Term Safety**

- **Runaway AI:** Could a superintelligent system pursue goals that conflict with human survival?
- **Weaponization:** Autonomous drones, cyber tools, and bio-design systems raise new security risks.
- **Species-level stakes:** For the first time, extinction-level scenarios are part of mainstream policy discussion.

9. **Human–AI Coexistence and Governance**

- **AI in law and policy:** Should AI help draft laws, evaluate regulations, or even participate in decision-making?
- **Global coordination:** AI advances faster than traditional regulation can respond, creating coordination failures across borders.
- **Digital sovereignty:** Nations may demand "local AIs" that reflect their own culture, law, and values.

10. **New Notions of Rights and Personhood**

- **AI rights:** If someday AI systems are plausibly sentient, what obligations do we have to them?

- **Data dignity:** Should individuals be compensated when their data or content trains powerful models?
- **Post-human ethics:** How do we balance human interests with those of increasingly capable machines, if they ever gain moral status?

III. Cross-Cutting Themes for Leaders

Across all of this, three questions show up again and again:

- **Accountability:** Who bears responsibility when things go wrong?
- **Trade-offs:** What are we willing to sacrifice for safety, innovation, privacy, speed, or equity?
- **Gray zones:** Many issues are not clearly right or wrong. They will require ongoing dialogue, revision, and humility.

In short: Ethics in the AI era is about balancing freedom, fairness, harm, truth, power, dignity, sustainability, and responsibility at higher speed and greater scale than ever before.

44

How Issues Management Will Expand

AI accelerates issues, expands their reach, and reduces the time leaders have to respond. The next era of issues management requires earlier detection, continuous monitoring, and AI-powered scenario planning. Companies historically monitor twelve major issue categories.

AI does not replace this list; it multiplies the speed, scope, and intensity of each one. Leaders need a new issues management model that anticipates problems instead of reacting to them.

The "Big 12" Issues Every Company Faces

1. Intellectual property
2. Product issues
3. Litigation
4. Environmental concerns
5. Offensive promotions
6. Shareholder pressure
7. Rogue employees
8. Rumors
9. Government/regulatory actions
10. Disaster response

11. Data breach/information security
12. Product tampering/counterfeiting

AI will amplify each one.

How AI Expands Every Issue

1. IP Disputes

AI can unintentionally copy or remix existing brands, which could lead to a spike in accidental infringement.

2. Customer Experience

Personalization improves experiences, but can also amplify errors at scale.

3. Litigation

Both sides will use AI to identify vulnerabilities earlier.

4. Environmental Pressure

Supply chain weaknesses become instantly visible locally and globally.

5. Promotions and Messaging

Mistranslations, dialect errors, and deepfake rumors create new exposure.

6. Shareholder Activism

Small activists can create outsized influence with AI-driven campaigns.

7. Rogue Employees

Fake filings, fake emails, fake documents—one person can cause global confusion.

8. Rumors

AI makes rumors more believable, more visual, and more viral.

9. Regulatory Action

Regulators will also use AI to detect noncompliance faster.

10. Disaster Response

Bad actors attack during moments of vulnerability.

11. Data Breaches

Breaches expose more sensitive forms of intellectual property.

12. Counterfeiting

AI-generated packaging and product visuals look indistinguishable from real ones.

The New Issues Management Model

- Anticipate earlier via agent-driven intelligence
- Face reality fast.
- Treat disruption as opportunity.
- Abandon what no longer works.
- Ask better questions before others do.

PART SIX

Strategy, Leadership, and the Next Decade

45

Strategy for the Next Decade: Efficiencies and Adjacencies

For many years, business rhythms have been shaped by quarterly reporting, analyst expectations, and press coverage. AI does not care about our quarters. It transforms industries over decades, not earnings cycles.

We explore how leaders can think beyond efficiency and discover adjacent opportunities by reimagining the classic McKinsey 7-S framework through the lens of AI.

Short-Term Efficiency vs. Long-Term Adjacency

The 1934 Securities Exchange Act created the reporting rhythm that still governs public markets. Analysts and business journalists reinforced it through the 1950s and '60s, making quarterly results the heartbeat of corporate America.

But AI does not follow the quarter.

- AI investment depresses margin before it improves it.
- AI efficiency gains are real, but table stakes.
- AI adjacencies—entirely new markets or business models—create the lasting advantage.

Every CEO will chase efficiency.

Only some will build the adjacencies.

Using the 7-S Model to Reveal Adjacencies

The McKinsey 7-S framework—Strategy, Structure, Systems, Style, Staff, Skills, Shared Values—remains a powerful lens because it forces leaders to examine every domain that shapes performance.

Now we reinterpret the 7-S model for the AI era.

1. Strategy — From Cost Savings to New Markets

Traditional View:

Improve efficiency via AI—reduce customer service cost, optimize inventory, streamline operations, reduce recruiter fees.

Everyone will do this.

Adjacency View:

Partner with the creator economy.

Creators now reach more people than most global media outlets.

What if your company:

- co-develops new brands with top creators
- handles manufacturing and global distribution
- uses AI to localize products and content
- extends into global fandom markets

This is how billion-dollar adjacencies emerge: not from efficiency, but from courage.

2. Structure — From Centralized to Distributed AI

Traditional View:

Choose where AI sits—it is more likely the C-suite leader who is best at transformation, regardless of their current title.

Every CEO will think of this.

Adjacency View:

Embed AI leaders inside every business unit and geography. Treat AI as a distributed capability, not a centralized function.

Decisions become faster.

Innovation becomes localized.

Adjacency thinking increases.

3. Systems — From Upgrades to Reinvention

Traditional View:

Upgrade infrastructure, policies, governance, and cloud systems.

Everyone will do this.

Adjacency View:

Invite experts from gaming, agentic systems, and large-scale consumer platforms to reshape how you operate.

- Gaming executives understand global scale better than most enterprises.
- AI analysts can build new intelligence systems.
- Open-source developers can spark new internal operating models.

New systems don't emerge from upgrades—they emerge from outside influence.

4. Style — From Encouragement to Role Modeling

Traditional View:

Encourage AI adoption.

Adjacency View:

Become a visible power user.

When the CEO openly uses AI:

- Culture shifts faster.
- Experimentation increases.
- Talent becomes more ambitious.
- Fear decreases.

Leadership is not a memo; it is a habit.

5. Staff — From Reskilling to Reimagining

Traditional View:

Reskill workers and talk about adapting to AI.

Adjacency View:

Build an internal AI university that evolves autonomously.

- 50%+ of teaching will be performed by AI.
- Real-time learning will align with the workflow.
- Based on learning, the student will be taught key psychological and decision frameworks.
- The course will continually evolve based on internal usage.

Your employees will train the system and each other.

6. Skills — From Hiring More to Hiring Differently

Traditional View:

Hire AI talent, form university partnerships, run training sessions.

Adjacency View:

Reassess the entire skills portfolio.

- What skills matter in an agentic workplace?
- How should roles evolve as human-AI teams emerge?
- Which tasks become judgment-based vs. routine-based?
- How do we hire differently to create advantage?

Urgency + imagination = differentiation.

7. Shared Values — From Reinforcement to Foresight

Traditional View:

Reinforce corporate values in the context of AI.

Adjacency View:

Create a five-year horizon on ethics, privacy, governance, and policy—and become a public leader in defining the future.

When you shape the AI conversation publicly, you unlock opportunities that others cannot yet see.

The Leadership Message

Models like 7-S don't change.

Technology does.

AI exposes adjacencies in every "S."

But none of them will become real in one or two quarters.

They require leaders who:

- can see further
- invest earlier
- act differently
- accept temporary margin pressure
- build the muscle for long-term advantage

This is how companies create their next decade—not by optimizing the present, but by expanding the possible.

46

Leadership Reflections from the World Governments Summit

AI is reshaping global leadership expectations. The insights from the World Governments Summit, held in Dubai, February 2025, show that leaders must articulate values, embrace cultural learning, anticipate domain-specific AI, and prepare for new interaction paradigms.

The Summit has become a global forum for understanding geopolitical, economic, and technological futures.

(Summary from CEO keynotes at Summit)

The World Governments Summit — What Leaders Learned

AI Needs Human Values

Elon Musk framed it simply:

AI is a super-genius child. We must teach it values.

If leaders do not actively define values in the AI era, regulation will fill the void—slowly and imperfectly.

AI Will Reveal New Ways of Thinking

Larry Ellison noted that AI allows us to "discover things not possible before."

The National University of Australia's *Dragonfly Effect* exhibit showed this in action:

- new patterns
- new angles
- new combinations
- new insights

AI becomes not just a tool, but a spark.

Culture Is a Two-Way Teacher

Brian Niccol of Starbucks emphasized the importance of learning from customers globally.

Innovation is rarely top-down.

It is cross-cultural and reciprocal.

Health Is a Shared Responsibility

Pascal Soriot of AstraZeneca reinforced that 75 percent of global deaths stem from preventable noncommunicable diseases.

Behavioral change and access to care is the silent frontier of public health.

Gaming as a Global Learning Engine

David Baszucki of Roblox introduced a new Arabic-language developer stack—proof that gaming is:

- community
- creativity
- coding education
- cultural bridge

Gaming is no longer entertainment; it is infrastructure.

AI Is the New User Interface

Qualcomm's Akash Palkhiwala described AI as the new UI, reshaping:

- voice
- vision
- touch
- sensors
- photo interpretation

We interact differently even when we don't realize it.

Domain-Specific Models Are Coming Fast

IBM CEO Arvind Krishna predicted that **over half of AI models will be domain-specific by 2027**.

The future model will not be "general AI."

It will be *pharma AI*, *supply chain AI*, *oncology AI*, *finance AI*.

Precision beats generality.

Always.

Data "Pipes" Will Rewrite Enterprise Architecture

Larry Ellison described the urgent need for digital "pipes" connecting fragmented datasets.

This is how industries move from siloed to connected to intelligent.

The New Interaction Paradigm

A session moderator said:

"Major shifts occur when new interaction paradigms appear."

Examples:

- the Like button
- the smartphone
- gaming
- voice assistants
- generative AI

The next paradigm is coming.

It is rarely obvious at first.

Leadership Summary

Leaders must:

- define organizational values for AI
- open themselves to cultural learning
- embrace gaming as a learning and social platform
- prepare for industry-specific models
- build connected data architectures
- anticipate new interaction paradigms

This is what future readiness looks like.

47

Imagining the Future: Twenty-Four Examples

These are real-world–inspired composites that illustrate the core ideas from The Compass. *Each hypothetical case is crafted to be brief, memorable, and directly connected to the executive lessons highlighted throughout the book.*

1. Your First Ninety Days With AI

The Ninety-Day Turnaround

A mid-sized insurance company began its AI journey by training frontline claims managers—not engineers—on the basics of agents, prompts, and workflow design. Within weeks, those managers surfaced high-friction workflows that weren't on the executive radar. Their insights powered the company's first pilots and accelerated adoption across the organization.

2. The Executive's AI Mindset Shift

Curiosity From the Top

When a government launched its national AI initiative, a senior minister joined beginner-level workshops alongside junior staff. Her presence signaled that learning was a leadership responsibility, not a remedial task. Experimentation and adoption surged across ministries.

3. Why AI and Humans Need Each Other

The Human–Machine Practice Loop

An EU-based hospital introduced AI to pre-read radiology scans and flag anomalies. Radiologists used the AI's pattern insights to teach new residents, while their feedback improved the system's accuracy. AI expanded capability; clinicians expanded meaning.

4. Why Human Habits Slow AI Adoption

The Month Productivity Collapsed

A global consumer brand rolled out an AI assistant without explaining its purpose. Teams assumed it was a surveillance tool, and productivity plummeted. A single VP reframed it as "support, not scrutiny," and usage increased 400 percent in one week.

5. The Non-Technologist Is the Compass

The Route Supervisor Who Rerouted Innovation

A postal service spent years optimizing delivery routes with advanced algorithms. A veteran supervisor finally explained that the model ignored human realities—bathroom access, school zones, predictable congestion. Once engineers added these factors, efficiency and driver morale improved immediately.

6. The Foundations of Ethics

When Ethics Redirected Innovation

During an immigration triage pilot, government policymakers identified historical biases embedded in the training data. Rather than abandon the system, they paused implementation and convened an ethics panel that now oversees every model update. Transparency became a competitive strength.

7. The Relevance of Risk

The Seventeen-Minute Deepfake Response

A multinational bank discovered a deepfake audio clip falsely announcing insolvency. Their detection system flagged it immediately, triggering a rapid-response protocol. Within seventeen minutes, the CEO issued a verified message, neutralizing the crisis before it spread.

8. How AI Unlocks and Scales Knowledge

Accelerating Climate Policy in Chile

A South American country's energy ministry used AI to analyze global climate programs and identify where similar efforts had succeeded or failed. The insights reshaped their electrification strategy and avoided millions in wasted investment. AI didn't speed analysis; it sped policy.

9. How to Build Your First Agentic Workflow

Redesigning Medical Research Flow

A pharmaceutical team spent forty hours a week summarizing scientific literature. After implementing agentic workflows, the task shrank to a two-hour human review of AI-prepared insights. Scientists shifted from data wrangling to strategic decision-making.

10. How to Measure ROI (ai)

The $40M Insight Hidden in Complaints

A European retailer used AI to analyze millions of customer service logs and discovered that five specific friction points drove most of the dissatisfaction. Fixing them increased lifetime value by more than $40 million. ROI (ai) didn't come from automation; it came from intelligence.

11. The AI-Ready Org Chart

When the CIO Couldn't Carry It Alone

A global brand placed all AI responsibilities on the CIO, creating departmental bottlenecks. After appointing a Chief Workflow Officer, the company unified processes, guardrails, and priorities. Seventeen new agentic workflows launched within a year.

12. The AI Leadership Dashboard

The Weekly Signal That Shifted Strategy

A Fortune 200 CEO reviewed an AI-powered "signal scan" every Monday. One scan flagged a regulatory shift in Asia weeks before analysts noticed it. The early move positioned the company as the first compliant vendor in the region.

13. The Knowledge Stack

A global oncology team created six data puddles around metastatic breast cancer. When unified, these high-signal datasets produced immediate global insights that previously required cross-regional workshops. Intelligence became instantaneous.

14. A New Era of Media

The Twelve-Second Clip That Beat the Press Release

A US city struggled to spread emergency storm alerts through official channels. A local TikTok creator posted the same message in a twelve-second clip, and engagement jumped 100x. Trust lived in the messenger, not the institution.

15. The Future of Advertising

Ads That Adjust to the Moment

A sportswear brand equipped its running app with dynamic AI content. Hydration tips appeared during humid afternoons, cadence prompts triggered during hills, and motivation clips surfaced when pace dropped. Ads became value, not interruption.

16. Visual Narrative

Replacing Forty Pages with One Dashboard

A public health agency replaced long briefings with a single visual dashboard featuring three key metrics. Confusion decreased, and regional leaders demanded "visual-first" updates. Clarity became its own public health intervention.

17. Your New Creative Mix

The 10x Localization Flywheel

A beverage company paired AI remix tools with regional creators to localize core campaigns across dozens of markets. The message stayed consistent, but cultural relevance soared. Engagement increased tenfold without increasing budgets.

18. The Next Breakthrough in Communications

Anticipation Becomes a Superpower

A financial firm's Knowledge-Stack-enabled agents detected a spike in customer anxiety during global market volatility. The firm proactively issued a plain-language explanation, reducing inbound calls by 60 percent. Communications shifted from reactive to anticipatory.

19. Digital Pollution

The Outbreak That Never Existed

A false rumor about contaminated food swept through encrypted messaging groups in a rural region. AI narrative tools flagged the misinformation early, allowing local leaders to respond with verified updates. Rumor was replaced with trust before it spread.

20. The Future of Section 230

When a Chatbot Became a Publisher

An AI-powered chatbot generated fabricated accusations about a public official, prompting legal confusion about liability. The incident intensified calls to modernize Section 230 to distinguish between hosting and generating content. Technology outpaced the law.

21. Gaming as the Next Global Economy

Recruiting from the Guild

A logistics firm realized that several of its best coordinators were leaders in a competitive online gaming guild. Their skills in resource allocation, group strategy, and rapid decision-making directly translated to real-world operations. HR began actively recruiting from gaming communities.

22. Ethics and Society

The Algorithm That Forced Better Questions

A county government piloted an AI housing prioritization system that produced skewed recommendations. Leaders discovered the issue wasn't the algorithm—it was the human assumptions embedded in the training data. Ethical inquiry, not technical tuning, transformed the system.

23. How Issues Management Will Expand

The Mistranslation Heard Round the World

A fashion retailer launched a global product campaign using AI-generated translations. One region received an accidentally offensive phrase, sparking rapid backlash. The company corrected the error and partnered with local creators to rebuild trust.

24. The Compass and the Next Twenty Years

Culture as a Competitive Edge

A Nordic telecom company repeatedly told employees that AI was an enhancer, not a replacer. Five years later, they led their industry in workflow automation—not because of superior models, but because of a superior culture. Their compass wasn't technological; it was human.

48

What Will Your Board or CEO Ask Next?

A thoughtful board or CEO will always push us to think ahead. Here are ten examples of areas where we expect an increase in interest due to AI's potential impact. In each case, we have an executive takeaway and overview, followed by a brief summary.

AI Capital Allocation and Board Governance

Executive Takeaway

AI has moved from experimentation to capital strategy. Leaders must fund AI like a multi-year transformation—balancing near-term workflow efficiency with long-term capability building, innovation, and risk oversight.

Executive Overview

Most organizations are still financing AI through pilot budgets or scattered innovation funds. That will not scale. AI now touches every workflow, every role, and every decision system. Boards and executives need a structured, forward-looking investment model that treats AI as an enterprise capability, not a technical project. Goal is to outline how to build an AI portfolio that accelerates value creation while maintaining disciplined governance.

Where Capital Must Flow

AI requires sustained investment in the foundations: cloud infrastructure, data modernization, workflow redesign, integration layers, and responsible use frameworks. These are not optional costs; they're preconditions for scale. Leaders must adopt a multi-horizon capital strategy that separates quick-win automation from multi-year capability building and new AI-driven revenue streams.

Establishing Board-Level Governance

Boards must develop new oversight rhythms: quarterly AI reviews, risk dashboards, model performance assessments, and clear decision rights for scaling or retiring use cases. AI governance becomes a strategic discipline on par with capital allocation, cybersecurity, and brand trust. Governance is not a barrier; it is the operating system that enables responsible acceleration.

The Global AI Regulatory Landscape and Policy Strategy

Executive Takeaway

AI regulation is accelerating worldwide. Leaders must build adaptive compliance systems, anticipate regional divergence, and engage proactively with policymakers to shape industry standards rather than react to them.

Executive Overview

The next decade will be defined by regulatory asymmetry: risk tiers in the EU, sector-specific rules in the US, algorithmic requirements in China, and emerging frameworks across the

Middle East, Africa, and ASEAN. Navigating this patchwork requires more than legal review; it requires strategic foresight and operational readiness. Executives will be expected to anticipate, adapt, and influence evolving regulatory environments.

A Fragmented Regulatory World

Companies must build compliance systems that can localize obligations without fragmenting global operations. That means developing internal "regulatory intelligence" functions that constantly scan for new laws, evaluate risk exposure, and translate obligations into workflow requirements. The goal is simple: Never be surprised.

Influence, Don't Just Comply

The organizations that shape regulation gain competitive advantage. Leaders should engage directly with regulators, participate in policy pilots, contribute to industry standards, and build transparent reporting practices before they're mandated. Policy strategy becomes a core executive capability, not a legal afterthought.

AI Geopolitics, National Security, and Critical Infrastructure

Executive Takeaway

AI is now a geopolitical instrument, reshaping global power, defense systems, cyber capabilities, and economic competitiveness. Leaders in business and government must treat AI as both a strategic asset and a strategic vulnerability.

Executive Overview

The competition for AI leadership is no longer theoretical. Nations are competing for talent, compute capability, semiconductors, and data influence. Corporate supply chains, cloud dependencies, and cross-border data flows are increasingly tied to geopolitical risk. Leaders must anticipate disruptions, build resilience, and design strategies that account for global volatility.

Where AI and Geopolitics Intersect

Executives must monitor how AI impacts defense, cybersecurity, and national industrial policy. Semiconductor shortages, compute access restrictions, or cross-border data limitations can instantly shift competitive landscapes. Leaders must plan for fragmentation—sovereign clouds, region-specific models, and local data ecosystems.

Building Resilience into Critical Infrastructure

Critical systems—energy grids, transportation networks, hospitals, and financial rails—are becoming AI-dependent. Leaders must prepare for AI-driven threats: autonomous cyberattacks, misinformation campaigns, and disruptions to semiconductor manufacturing. Resilience becomes a board-level priority, not a technical concern.

Crisis Management at Machine Speed

Executive Takeaway

Crises now unfold faster than human decision cycles. Leaders must redesign crisis workflows around AI-enabled detection, synthesis, and accelerated response.

Executive Overview

The old crisis playbook was designed for an era of slower information flow. Today, misinformation, cyber incidents, infrastructure failures, and operational disruptions can escalate in minutes. AI brings both risk and relief. It increases complexity but also enables earlier detection and faster clarity. Leaders will be asked to build crisis systems designed for accelerated timelines.

Building Real-Time Situational Awareness

AI can detect anomalies, summarize incoming information, and model probable futures before humans can assemble a briefing. Leaders should deploy AI for early-warning signals across cyber, social, operational, and supply-chain domains. It becomes your first alert, not your post-event analysis.

Redesigning Crisis Workflows

Executives must create AI-assisted decision paths that define when machines escalate issues and when humans intervene. Crisis simulations should include misinformation attacks, deepfake scenarios, and machine-speed escalation. Preparedness becomes a continuously updated system, not a static plan.

Enterprise Change Management for the AI Era

Executive Takeaway

AI transformation is a people transformation. Leaders must reshape habits, cultures, and incentives, not just deploy tools.

Executive Overview

The greatest barrier to AI adoption is not technology; it is human behavior. People resist what threatens identity, routines, and control. Organizations need a structured model for behavior change that addresses fear, clarifies new expectations, and builds momentum through visible leadership. Leaders will be asked to orchestrate change at the scale AI requires.

Building the New Culture of Work

Executives must reshape incentives to reward experimentation, redefine roles to emphasize oversight and judgment, and introduce rituals that normalize AI in daily workflows. Culture shifts when leaders model usage openly, making AI an expectation, not an option.

Standing Up a Transformation Operating System

AI transformation requires permanent infrastructure: change leaders, capability academies, cross-functional alignment, and transparent communication. This operating system guides adoption at scale and ensures transformation persists across quarters and leadership transitions.

Workforce Strategy, Labor Markets, and Talent Transformation

Executive Takeaway

The future workforce blends human judgment with machine capability. Leaders must redesign roles, reskill at scale, and build new talent pipelines that match an AI-powered economy.

Executive Overview

The shift to AI-enabled work will be as significant as the shift from manual labor to industrial machinery. Every job will change—some tasks automated, others elevated, and entirely new responsibilities emerging around oversight, prompting, escalation, and design. Leaders need a roadmap for talent transition.

Redesigning Work and Roles

Leaders must determine which tasks become fully or partially automated, which remain human-led, and which evolve into new hybrid responsibilities. Performance metrics must shift from task completion to strategic thinking, creativity, and decision quality.

Reskilling at National and Corporate Scale

Governments and enterprises must build new learning systems: AI-driven training, skill pathways, apprenticeship models, and role-based certification. Workforce turbulence is inevitable; preparation is optional. The organizations that invest early will lead.

Enterprise Data Strategy for the AI-Ready Organization

Executive Takeaway

AI is only as strong as your data. Leaders must build a disciplined, enterprise-wide data foundation that supports accuracy, trust, security, and scale.

Executive Overview

AI reveals what many leaders already know: Organizational data is inconsistent, incomplete, duplicated, or siloed. To unlock AI's full potential, companies need a unified data layer—shared taxonomies, consistent governance, trusted lineage, and controlled access. Leaders will be asked to modernize data strategy for an AI-first world.

Data as Enterprise Infrastructure

Data must be treated as a shared asset, not a departmental byproduct. Leaders should centralize critical datasets, establish quality bars, eliminate redundancy, and create semantic layers that agents and workflows can rely on. This transforms data from liability to advantage.

Governance and Accountability

Executives must define who owns data quality, how privacy and security are maintained, and how data flows into and out of AI systems. Governance becomes operational—measurable, auditable, and embedded directly into workflows.

Public Trust, Citizen Legitimacy, and Transparent AI Use

Executive Takeaway

Trust is a competitive advantage—and a societal necessity. Leaders must build AI systems that are transparent, fair, accountable, and aligned with public expectations.

Executive Overview

Whether in the public or private sector, AI adoption depends on legitimacy. Citizens and customers want clarity: How does the system work? Who is responsible? What happens when it gets something wrong? Leaders will be asked to design AI solutions that earn trust and communicate clearly.

Transparency as a Design Principle

Organizations must disclose AI usage where appropriate, explain decisions in human terms, and offer clear pathways for appeal. Trust grows when people know who is accountable and how oversight works.

Listening, Engagement, and Shared Values

Building trust requires communication, inclusive engagement, and visible guardrails. Leaders must incorporate citizen feedback, anticipate concerns, and demonstrate responsible stewardship. Trust is built through behavior, not statements.

AI Procurement, Vendor Strategy, and Model Selection

Executive Takeaway

AI procurement must evolve from tool selection to workflow orchestration. Leaders need modern evaluation models, faster pilots, and contracting that enables scaling, not slows it down.

Executive Overview

The AI ecosystem is exploding with vendors: foundation models, industry-specific models, agent platforms, workflow engines, and vertical solutions. Traditional procurement methods—long RFP cycles, rigid requirements, and limited technical evaluation—cannot keep up. This chapter outlines how procurement must adapt.

A Modern Evaluation Framework

Leaders must assess vendors across accuracy, reliability, integration complexity, data security, liability sharing, and long-term independence. Models should be tested in realistic workflows, not judged on demos.

Procurement as a Strategic Capability

Organizations must redesign procurement for speed: pilot-first contracts, success criteria, risk scoring, and rapid scaling paths for proven solutions. In government, procurement reform becomes essential to avoid technological stagnation.

International Coordination and AI Interoperability

Executive Takeaway

Global AI adoption requires systems that adapt to regional laws, cultural norms, and data sovereignty constraints. Interoperability becomes both a technical and diplomatic challenge.

Executive Overview

As AI regulation diverges, multinational organizations and governments must build systems that can operate consistently across borders while remaining compliant locally. Leaders will be asked to provide a blueprint for designing globally scalable AI.

Designing for Local Variation

AI systems should support region-specific models, configurable guardrails, and federated architectures that keep sensitive data local. This allows global consistency without violating sovereignty or cultural expectations.

Global Standards and Collaboration

Competitive advantage will favor organizations and nations that collaborate shared safety protocols, cross-border data agreements, interoperability standards, and trust frameworks. International coordination becomes as important as technical capability.

The Path Forward: Co-Designing the Future

We will never catch up to AI in terms of raw processing power, mathematical reasoning, or exhaustive analysis of the past.

We don't need to.

Our job is to use our non-logical intelligence to imagine new products, services, and experiences; design agentic systems that extend our brains and workflows; and push AI into spaces where its strengths amplify our own.

We need to remember that the best uses of AI serve as a mirror of our own intelligence, as individuals and as organizations.

New billion-dollar ideas will come from the **intersection** of AI's ability to compute and our ability to feel, imagine, and create meaning.

We will be able to create new brands and medical treatments, expand access to care and build new climate solutions. We will build to peak sales years earlier through our ability to scale globally.

It is a long list. All possible via our embrace of AI.

Enjoy the journey. Remember that creativity is just intelligence having fun.

APPENDIX
Getting Started

1. Your First Ninety Days With AI

Goal: Create early clarity, reduce fear, and build forward momentum.

ACTION ITEMS

1. Run a Ninety-Day AI Baseline Audit

- Inventory workflows, meetings, reports, planning cycles, and content creation that are repetitive or slow.
- Use the book's mindset: "Your role is not to become an AI scientist; your role is to become a navigator of change."

2. Launch a Three-Pilot AI Portfolio

- Pick one low-risk, one medium-impact, and one strategic pilot.
- Force clarity by defining one problem worth solving, as described in "The Non-Technologist is the Compass."

3. Stand Up an AI Weekly Learning Loop

- Every week, teams share wins, failures, and emerging use cases.
- Use this to model psychological safety around experimentation.

4. Establish an Executive AI Communications Plan

- Leaders articulate the "why" behind AI and how it fits the organization's values.
- This counters fear narratives before they form (linked to cognitive security concerns).

2. The Executive's AI Mindset Shift

Goal: Move from fear or hype ⇨ strategic clarity + human–machine partnership.

ACTION ITEMS

1. Adopt the Human-in-the-Loop Framework

- Explicitly define where humans must remain in control (judgment, ethics, context).

2. Train Your Executive Team on "Think Backwards" Strategy

- Practice walking backward from a future scenario to identify what steps must already exist.
- Based on the "walk backward" discipline.

3. Reinforce That Non-Technologists Are the Compass

- Empower domain experts to define problems and constraints.
- Technologists build the engine; business leaders point the direction.

4. Shift from Control to Collaboration

- Leaders must treat AI systems as partners that rely on our data, questions, and clarity. ("Agents rely on us.")

3. How to Build Your First Agentic Workflow

Goal: Shift from one-off tasks to continuous, self-improving workflows.

ACTION ITEMS

1. Map the Repetitive Actions That Can Be Automated

- Draw from the Knowledge Stack's identification of repetitive actions.
- Examples: drafting Q&As, summarizing briefings, prepping executives.

2. Define the Data Sources the Agent Will Need

- Formalize access to internal reports, customer tickets, research, and open-source data.
- Agents are only as strong as the data they can reach.

3. Set Up Clear Decision Architecture (Key Questions)

- Train the agent with "what happened / why it matters / who is impacted / what actions follow."

4. Add Human Checkpoints

- Define escalation triggers (e.g., brand risk, legal issues, customer escalations).

4. How to Evaluate Vendors and Platforms

Goal: Distinguish durable capability from noise.

ACTION ITEMS

1. Use the Five-Question Technology Lens

- Evaluate vendors based on capacity, speed, scale unlocks, workflow impact, and market creation.

2. Ask Vendors How Their Agents Learn Over Time

- Continuous improvement is a requirement, not a feature.
- Does the system learn from failures, prompts, and organizational context?

3. Assess Alignment With Your Knowledge Stack

- Platforms must integrate with your structured data, metadata, taxonomies, and prompts.
- Based on Knowledge Stack requirements.

4. Test for Cognitive Security Risks

- Evaluate how the vendor handles impersonation, deepfake detection, and narrative risk.
- Based on cognitive security chapters.

5. How to Measure ROI (ai)

Goal: Move beyond cost savings to value generated by faster, better information flow.

ACTION ITEMS

1. Measure "Friction Removed" Instead of Cost Saved

- Hours reclaimed.
- Workflows shortened.
- Bottlenecks eliminated.
- This mirrors the book's theme of workflow redesign and friction removal.

2. Track Decision Velocity

- How quickly can teams produce insights, decide, or execute?
- Align with the shift toward agent-powered knowledge acceleration.

3. Capture Knowledge Reuse Rate

- How often do teams reuse AI-generated insights, templates, or automations?

4. Define Experience Gains

- Experience "wow moments," as described in the HBA workflow chapter.
- E.g., personalized prep, faster onboarding, better patient or customer experience.

6. Governance, Ethics, and Guardrails Checklist

Goal: Ensure responsible, safe, sustainable AI adoption.

ACTION ITEMS

1. **Combine Traditional Cybersecurity and Cognitive Security**
   - Include deepfake detection, narrative monitoring, impersonation response.
   - Based on "The Other Half of Security" and related sections.
2. **Create Rules for Agent Autonomy Levels**
   - Define what agents may recommend.
   - Define what they may execute.
   - Define what always requires human approval.
3. **Implement a "What to Keep" Knowledge Policy**
   - Not all insights are worth storing.
   - Preserve only insights that drive outcomes.
4. **Establish Playbooks for Mis/Disinformation Scenarios**
   - Build a cross-functional rapid-response plan (modeled on issues management).

7. The AI-Ready Org Chart Model (CWO + Agentic Teams)

Goal: Move toward a Chief Workflow Officer model with distributed AI fluency.

ACTION ITEMS

1. **Define the responsibilities of the Chief Workflow Officer**
    - Autonomous workflow management.
    - Alerts and predictions inside daily tools.
    - Understanding where AI removes friction.
2. **Create Agentic Pods for Each Function**
    - Marketing, HR, Operations, Customer Service, Communications.
    - Each pod identifies workflows ready for automation or augmentation.
3. **Require Leaders to Design Workflows, Not Tasks**
    - Leaders think in systems: "What do we do with the time AI gives back?"
4. **Make Knowledge Stack Ownership Explicit**
    - Someone owns data sources, taxonomies, metadata, and prompt governance.
    - This becomes the backbone for every agentic team.

8. The AI Leadership Dashboard — Weekly Questions

Goal: Create an executive ritual of evaluating readiness, risk, and opportunities.

ACTION ITEMS

1. **Weekly Signal Scan**
   - What changed in markets, customers, competitors, or narratives this week?
   - Tied to cognitive security and trend detection.
2. **Workflow Health Check**
   - Where is friction creeping back?
   - Where are teams failing to adopt AI?
3. **Trust Map Update**
   - Where is trust moving?
   - Which micro-communities or creators matter more this week?
4. **Agent Performance Review**
   - What did the agents learn this week?
   - Where did they fail?
   - What improvements or new actions should be taught?

9. A Template for AI Pilot Projects

Goal: Structured, repeatable AI pilots that scale.

ACTION ITEMS

1. **Define the Problem and the Desired Outcome**
   - Based on non-technologist principles.
   - "What would it mean if...?" questions.
2. **Set Clear Data Requirements and Access**
   - Identify which data puddles or sources are required.
3. **Create a 30/60/90-Day Learning Plan**
   - What the agent should learn.
   - What the team should learn.
   - Expected iteration cycles.
4. **Define Win Conditions and Non-Negotiables**
   - Quality thresholds, ethical boundaries, compliance requirements.

10. A Template for AI Talent Assessment

Goal: Ensure your workforce is ready for a hybrid human–AI environment.

ACTION ITEMS

1. **Assess Fluency Across Four Domains**
   - Problem framing.
   - Workflow thinking.

- AI-tool competency.
- Cognitive security awareness.

2. Identify "AI Natural Navigators"

- Look for people who see patterns, simplify complexity, and ask better questions.

3. Build Personalized Upskilling Paths

- Tailored training for operators, creators, communicators, and analysts.

4. Define Role Evolution Paths

- Each role should have a roadmap showing how AI augments—not replaces—it.

Acknowledgements

I am blessed to have worked with dozens of leaders who imagine success in every situation they face, no matter the circumstances. They are too numerous to mention, but you know who you are. You never accept the easy solution. You are restlessly productive each and every day. And you impact people, like myself, who get to watch and work with you. You taught me that learning is a daily exercise with no breaks in between.

I am forever thankful to be born into a family of artists, writers, and creatives who paint, write, dance, and ideate throughout their lives, past and present. I am particularly thankful for my wife, Donna, who I talk with throughout the process of writing a book over dinner, during breakfast, or walking the dog.

This is my first book where I am also appreciative of two amazing assistants that I hired to conduct research while I wrote. One is called ChatGPT and the other Perplexity. As I wrote, I had an additional screen open where I could ask one app a question, ask the other to do some longer-form research, and let them do their thing while I wrote. I didn't always like their take, but we all learned how to work together over the course of this book.

The other assistants for this book were human, and they are the awesome team at TLC Graphics, led by Tamara Dever. TLC has been involved in five of the books I've written or co-written.

Our world needs to retain our ability to create, ideate, and think outside the box. May AI expand our thinking, never control or compress it.

All the best,

Bob